WHAT KIND OF ROPE—WHAT KIND OF SPLICE ?

ANY ROPE

Does the rope have a core? See Rope Know-How/ Construction, page 9

NO

Does the rope consist of high-modulus fiber e.g. Dyneema, Spectra, etc.? See Rope Know-How / Material, page 12

NO

or only in part

Chapter 1.1
Eye / thimble splice in conventional and mixed fiber ropes
page 37

YES

Chapter 1.2
Eye / thimble splice in high-modulus fiber ropes
page 44

OR

Chapter 1.3
Joining rope splice
page 60

OR

Chapter 1.4 Loops
page 65

OR

Chapter 1.5
Soft shackles
page 75

OR

Chapter 1.6
Lashing /Transfilage
page 105

YES

Does the core consist of high-modulus fiber e.g. Dyneema, Spectra, etc.? See Rope Know-How / Material, page 12

NO

Is the core braided?

No, the core has parallel fibers or yarns

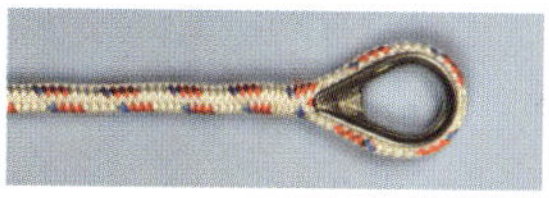

Chapter 2.2
Eye / thimble-splice in parallel core
page 128

YES

Is the core braided?

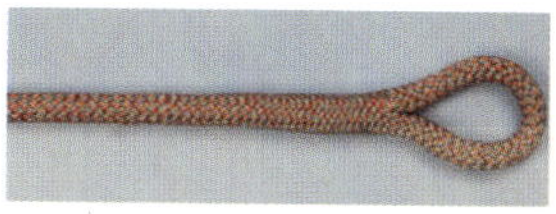

Chapter 2.1
Eye / thimble splice
page 110

OR

Chapter 2.3
End-to-end splice
page 136

OR

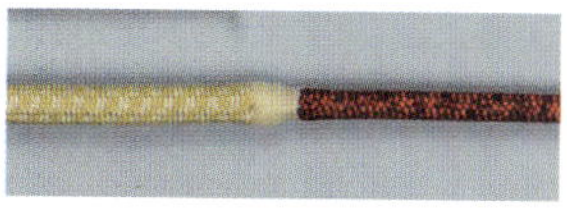

Chapter 2.6
Additional cover splice
page 187

YES

Chapter 2.4
Eye / thimble splice in high-modulus fiber ropes page 151

OR

Chapter 2.5
End-to-end splice In high-modulus fiber ropes
page 164

OR

Chapter 2.6
Tapering and additional cover splice
page 183

Modern SPLICING

EGMONT M. FRIEDL

SCHIFFER PUBLISHING

4880 Lower Valley Road • Atglen, PA 19310

Library of Congress Control Number: 2020943646

Cover design by Brenda McCallum
Type set in Gotham/Geoslab

Translated from the German by
Egmont M. Friedl

Originally published as *Modernes Tauwerk Spleissen*
© Delius Klasing & Co. KG, Bielefeld, 2019

Cover Images: Sport climbing, Stock Photo ID: 359812600 Courtesy of www.bigstockphoto.com. Spinnaker with ropes on deck of boat, Stock Photo ID: 277093273, Courtesy of www.bigstockphoto.com. Kids climb on high rope trail, Stock Photo ID: 289062031, Courtesy of www.bigstockphoto.com

ISBN: 978-0-7643-6181-4
Printed in India

Published by Schiffer Publishing, Ltd.
4880 Lower Valley Road, Atglen, PA 19310
Phone: (610) 593-1777; Fax: (610) 593-2002
E-mail: Info@schifferbooks.com
Web: www.schifferbooks.com

Modern SPLICING

CONTENTS

PREFACE

This book will demystify and guide you through all the important splicing techniques of modern braided cordage. All instructions are presented in great detail so that even complicated splices can be mastered successfully. Additionally, a DVD is included that shows all the splicing work presented in the book in equally clear and precise hands-on videos.

Reducing the contents to splices that are important in all practical purposes must not be confused with simplifying their execution or leaving out any of the more complicated techniques—quite to the contrary! Every type of splice is shown in the best practices—and that means the fully involved, professional method. What has been excluded are solely obsolete versions of inferior quality and strength. The chapter "Modern Rope Know-How—What to Know before Starting to Splice" similarly concentrates on what is important when actually working with and splicing ropes for practical purposes, and will give very solid understanding and competence of modern cordage, its construction, and fiber materials.

Splicing is and remains a craft, and as is the case with all manual work, skill comes with practice. It should nevertheless be possible to fabricate any of the splices presented in this book on the first try—presumed you›re working with reasonable care. It might be wise not to expect an absolutely perfect result right away, so best to practice on some spare, unused rope before cutting and splicing actual sheets and halyards. Working with one's hands and learning a craft can give tremendous satisfaction, especially in our times with such excess of digital and virtual occupations. Holding an accomplished, self-made splice in your hands—something only few know how to make—is a thing to be rightly proud of. Along these lines, I want to wish all readers and splicers best success and enjoyment with Modern Splicing.

Egmont M. Friedl
EMF Marine
Rigging and splicing tools
Online shop: www.emf-marine.com

INTRODUCTION

Why are lines spliced? There are two main reasons.

First of all, a splice gives the strongest way of attaching or joining the end of a rope to any object or to another rope. Every knot weakens a rope. Usually about 40% of the breaking strength of the rope is lost by attaching it with a knot. With modern high-modulus fiber ropes, the loss of tensile strength is even greater. Dyneema or Spectra will loose about 65% of their breaking strength if knotted. Sustaining a loss of more than half the breaking strength in a knotted termination is simply unacceptable. These ropes need to be spliced! A splice can retain the full breaking strength of the rope, but for safety reasons it is best practice to assume retaining 80% of breaking strength by a splice.

The second reason is that a proper splice just looks elegant, crafty, and seamanlike.

So it is no wonder that on the most pristine and elegant classic yachts as well as on board any modern high-performance regatta yacht, sheets, halyards, and control and mooring lines are all meticulously spliced.

Splicing braided and especially double-braided ropes may seem a little enigmatic to many, and quite often the option of taking such work in one's own hands is discarded without further consideration. I hope this book can help change that perception, because neither a specialized workshop nor equipment is necessary to produce a professional-grade splice. Often the time needed to splice a rope is quoted for how good someone can splice. While this may be important for the piece-working professional, for the sailor splicing his own running rigging it is of only minor concern, whether needing 10 or 20 minutes for one piece of work. The important thing is to achieve a proper result and a maximum-strength splice.

The biggest obstacle to success in splicing is to not take the work seriously enough. A good splice cannot be achieved when huddling or hurrying. That's a recipe for guaranteed frustration! And not only will it be far more successful when working with care and diligence, it will also be much more enjoyable. In countless splicing workshops I was able to observe how a little haste or impatience can deter from achieving good results. So my recommendation is this: Take your time, work with precision, and imagine yourself to be a surgeon, a rope surgeon so to speak, working on "open-rope surgery." And with that mindset there should be nothing in the way to success!

MODERN ROPE KNOW-HOW—WHAT TO KNOW BEFORE STARTING TO SPLICE

Braided ropes are specified first by their construction and second by the material the rope is made from.

BRAIDED-ROPE CONSTRUCTION

According to their construction, braided ropes can be split into two major categories: single-braided ropes and double-braided ropes. All the splicing instructions in this book are divided the same way, into two parts. The first part shows splicing of single-braided ropes, and the second part shows splicing of double-braided ropes.

Single-braided ropes, also called hollow braid, have no core inside the rope. Some of these ropes have a little string on the inside. This is not a core and serves only to give the rope a nice round section. This string is removed over the whole length of the splice. Single-braided ropes do not necessarily consist of only one type of fiber. If yarns of different fibers are braided parallel to each other into a single-braided rope, this becomes a mixed-fiber rope. A typical such mix could consist of yarns made out of HMPE (such as Dyneema) and yarns out of polyester.

Double-braided ropes consist of a core and a cover. This construction can also be called kernmantle rope. The cover is the outside braid; the core inside is usually also braided but may consist of parallel-lying loose fibers (see chapter 2.2) or it may consist of several twisted strands. It is important to know that core and cover are not affixed to each other. The core just lies inside the cover. This is why the core can be extracted at any given spot along the length of the rope (see the chapter "Basic Splicing Skills"). When splicing double-braided rope, core and cover are spliced separately. The construction of double-braided ropes has many advantages. One of the advantages is that different materials can be used for the core and the cover. This way the positive properties of one material can be used for the outside, and another material with its special attributes for the inside. Typically, a good-quality halyard consists of a low-stretch core made of high-modulus fibers such as Dyneema and a cover that is resistant to abrasion and is durable and UV resistant, such as polyester, which is also more economically priced. In such a rope, the load is carried by the core alone. The cover serves only as protection. This is why ropes with a high-modulus fiber core need to be spliced with a core-to-core splice (see chapter 2.4). Some double-braided ropes have an additional cover braid around the core, which is called an intersheath. This intersheath is a thin, hollow braid that lies between the actual core and the cover of the rope. Its function is to create friction between the core and the cover and avoid any slippage of the core, especially when the rope is locked in jammers or rope clutches. Over the length of the splice, the intersheath is removed. In ropes with high-modu-

Single-braided ropes have no core.

1 Polyester 12-strand or 12-plaited (gray)

2 Dyneema 12-strand or 12-plaited, coated (red)

3 Dyneema 12-strand or 12-plaited, coated (black)

4 Mixed-fiber, parallel yarns from Dyneema (white) and polyester (red), 8-strand or 8-plaited, small string inside (no core) to give nice, round section

5 Mixed fiber, strands consist of Dyneema (white) and polyester (green), 8-strand or 8-plaited, small string inside (no core) to give nice, round section

6 Mixed fiber, strands consist of Dyneema (white) and polyester (blue), 8-strand or 8-plaited

7 Mixed fiber, strands have a high ratio of 70% Dyneema (white) and 30% polyamide (black), 8-strand or 8-plaited

lus cores but without an intersheath, friction between core and cover is generated by a special coating of the core.

In contrast to ropes with high-modulus cores, a conventional double-braided rope with core and cover made of polyester or polyamide has the load distributed about equally between core and cover. This makes it necessary to splice the cover over a greater length into the core when working with double-braided ropes made of conventional fibers (see chapter 2.1).

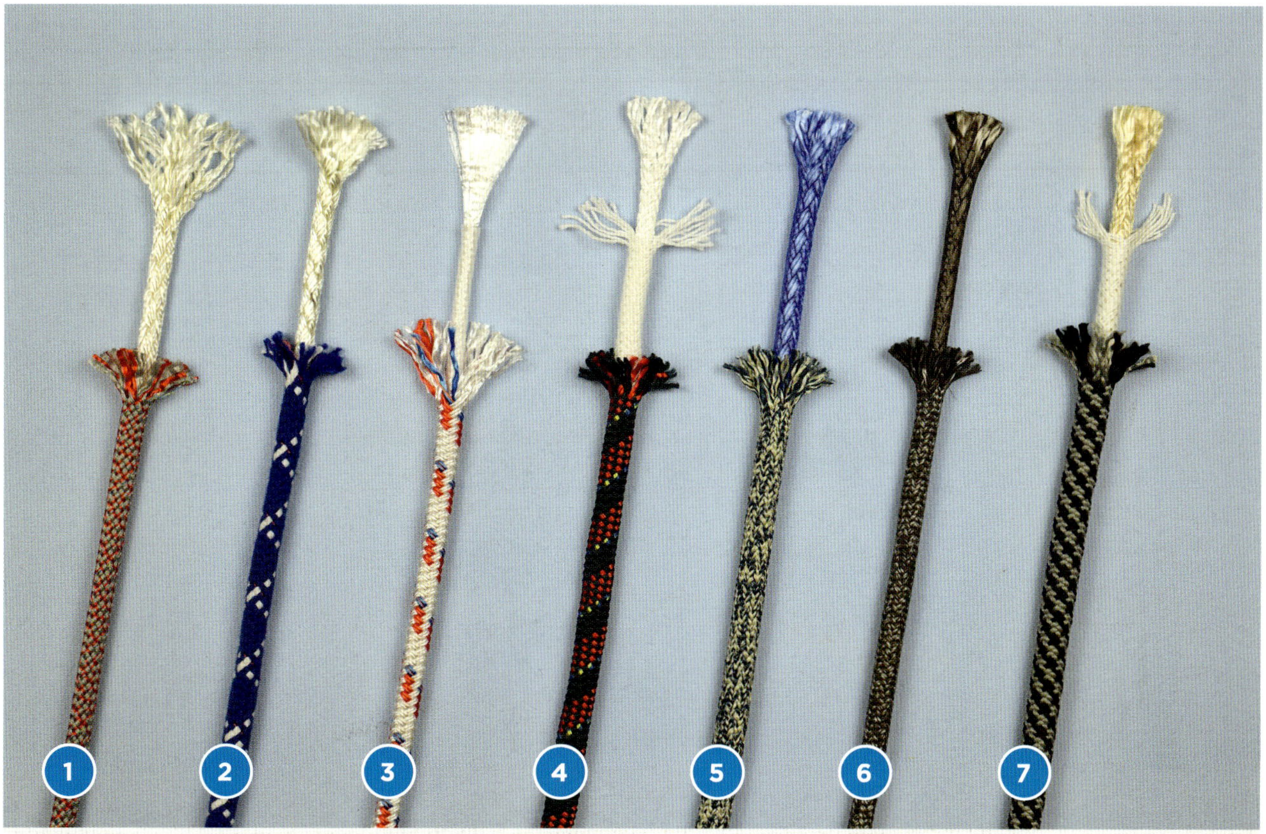

Double-braided ropes consist of a core and a cover.

1. Core from polyester, cover from polyester, 1:1 braid (red, silver)
2. Core from polyester, cover from fluffy polyester staple yarn, 1:1 braid (blue, white)
3. Core from parallel polyester fibers, with intersheath, cover from polyester, 2:2 braid (white, red, blue)
4. Core from Dyneema, noncoated, with intersheath, cover from polyester, 1:1 braid (dark blue, red, yellow)
5. Core from Dyneema, coated, no intersheath, cover from mixed fiber of Vectran and polyester 2:2 braid (dark blue, yellow)
6. Core from Dyneema, coated, no intersheath, cover from mixed fiber of Technora and polyester, 2:2 braid (gray, black)
7. Core from Vectran, noncoated, with intersheath, cover from polyester, 2:2 braid (black, silver)

The cover of a rope can be braided in different ways. In a 1:1 braid (one-over-one braid), each strand goes once over and once under, while in a 2:2 braid, each strand crosses over two opposing strands. A 1:1 braid is more resistant to abrasion and has a smoother surface and more stiffness. A 2:2 braid creates more grip and a more flexible rope. In addition to that, the braid is also specified by the number of strands. Typical braids are 12-, 16-, 20-, 24-, or 32-strand or plaited ropes.

1 2:2 braid, 12-strand or 12-plaited (blue, silver)

2 2:2 braid, 20-strand or 20-plaited (black, white)

3 1:1 braid, 16-strand or 16-plaited (black, blue, white)

4 1:1 braid, 32-strand or 32-plaited (red)

MODERN ROPE MATERIALS

In a second step after determining a rope's construction, modern braided ropes are specified by the material they are made from. Again, all braided ropes can be split into two major categories: ropes made out of conventional fibers and ropes made out of high-modulus fibers.

Conventional fibers are polyester, PES (trade names: Trevira, Dacron, Terylene, etc.); polyamide, PA (trade names: Nylon, Cordura, Perlon, Caprolan, etc.); and polypropylene, PP (traden ames: Hempex, Nelson, Ulstron, etc.). These conventional fibers are often blended together in order to combine positive attributes or to reduce cost. Such ropes are called copolymer ropes and are marketed under, again, different trade names (Deltaflex, Jetcore, Nystron, Supermix, Poly-Plus, etc.).

Polyester is very abrasion resistant and has very low water absorption, good strength, and durability. Polyester staple yarns can be used to produce a soft, fluffy cover braid with high grip for manually handled ropes. Polyamide, in contrast, has much more elasticity and markedly higher water absorption, with about equal breaking strength. It is more suitable for mooring lines and anchor rodes.

Polypropylene has significantly less breaking strength but floats in water, which makes it suitable for maritime rescue lines. Polypropylene is a comparatively low-cost material for rope production.

High-modulus fibers are extremely strong fibers that are used in rope production to yield incomparably higher breaking strength than conventional fibers. Additionally, high-modulus fibers have extremely low stretch characteristics. In principle, this increase in performance is achieved during synthesis by aligning the molecules parallel to each other, and thus exactly in the direction of the load. Given the same section, high-modulus fibers have much-higher tensile strength than steel at only a fraction of the weight. It is no wonder that these modern fibers revolutionized the usage of textile ropes. In countless applications, steel wire is replaced by high-modulus fiber ropes, such as for lifting gear in industry, in offshore oil and gas work, in commercial shipping, in forestry, and in the offroad vehicle sector. In the future, elevators may be suspended by textile ropes instead of steel wire, and it is even hard to say which branch of industry will not be affected by high-modulus fibers.

In marine sports the most important modern fiber is high-modulus polyethylene (HMPE) or ultra-high-molecular-weight polyethylene (UHMWPE), which is best known under the trade names of Dyneema or Spectra. Not only are these fibers extremely strong, they are also very durable, do not absorb water, are UV resistant, and even float in water. Their only drawback is a relativly low melting point of 140°C (284°F) and markedly low tenacity in knots, amounting to only about 35% of the breaking strength of the rope. This is the reason why Dyneema or Spectra lines should not be knotted, but always spliced! The originally white HMPE fibers used to be solely produced by the Dutch company DSM under the trade name of Dyneema, and by Honeywell in the United States under the trade name of Spectra, who ship their high-modulus yarns to rope manufacturers the world over. Nowadays, patents for HMPE fiber technology have expired, so that other names and producers have appeared on the market. It is only during production of an actual rope that the superior properties of high-modulus fibers are fully exploited by special coatings and thermal setting of the rope under very exacting tension and temperature. These processes allow the ratio of breaking strength to section to be maximized and decisively improve resistance to abrasion and the surface finish of the rope. Thus it is not only the fiber itself but also the manufacturing process that determines the final quality and performance of a high-modulus fiber rope. The coating simultaneously allows the yarns or the rope to be dyed to any desired color. Under high static load, most HMPE fibers creep slowly, elongating the length of a rope. When tension is released, the rope does not regain its original length. This inelastic deformation excludes common HMPE fibers from use in dimensionally stable applications such as standing rigging; for example, shrouds and stays on a sailing yacht. There are many different subtypes of HMPE fibers, and some of them even exhibit no creep. Research and development in high-modulus fibers remains very active to this day.

In the maritime sector, Dyneema and Spectra have largely superseded some other high-modulus fibers such as Kevlar. Fibers that still play an important part in the sailing and maritime world include LCP, trade name Vectran, which has a similarily high breaking strength as HMPE and does not exhibit creep under static strain but is not very resistant to UV radiation. Another high-end fiber is PBO, trade name Zylon, which has the highest breaking strength and exhibits no creep. This fiber can be used for standing rigging but must be carefully protected from UV radiation and moisture.

Further information on different fibers can be found toward the end of this book in the chapter "Technical Characteristics and Comparison Data on Fibers Used in Modern Ropes," on page 188.

HOW TO DETECT WHAT MATERIAL A ROPE IS MADE FROM

If the manufacturer and trade name of a rope are not known, it is not easy to discern the materials a rope is made from by looks or sensory touch alone. Most important will be to determine whether the rope consists of high-modulus fiber or conventional fiber, because the correct splicing technique also depends on it. With experience, noncoated Dyneema or Spectra can be made out by its silky, shiny appearance. Another way is to cut a short end off and test if it floats in water (only HMPE and polypropylene float). It›s best to hold the material into an open flame or to a heat cutter or soldering iron. Conventional fibers (PES, PA, PP) show dark edges when they melt. With longer exposure, they ignite and change into a liquid state. After cooling off, the melted area is hard with sharp edges. Dyneema and Spectra, in contrast, react to less heat in a very typical way. They pan out, giving the end of a rope a mushroom form, meaning the end distinctly expands in size under heat. With more heat, HMPE burns with a slightly blue flame. Vectran, Kevlar, and Zylon are of gold-yellow color if undyed and are very resistant to heat. They smolder and carbonize when held to an open flame, without actually igniting, while giving off small black charred particles. Do not breathe in any toxic fumes when heat-testing synthetic fibers!

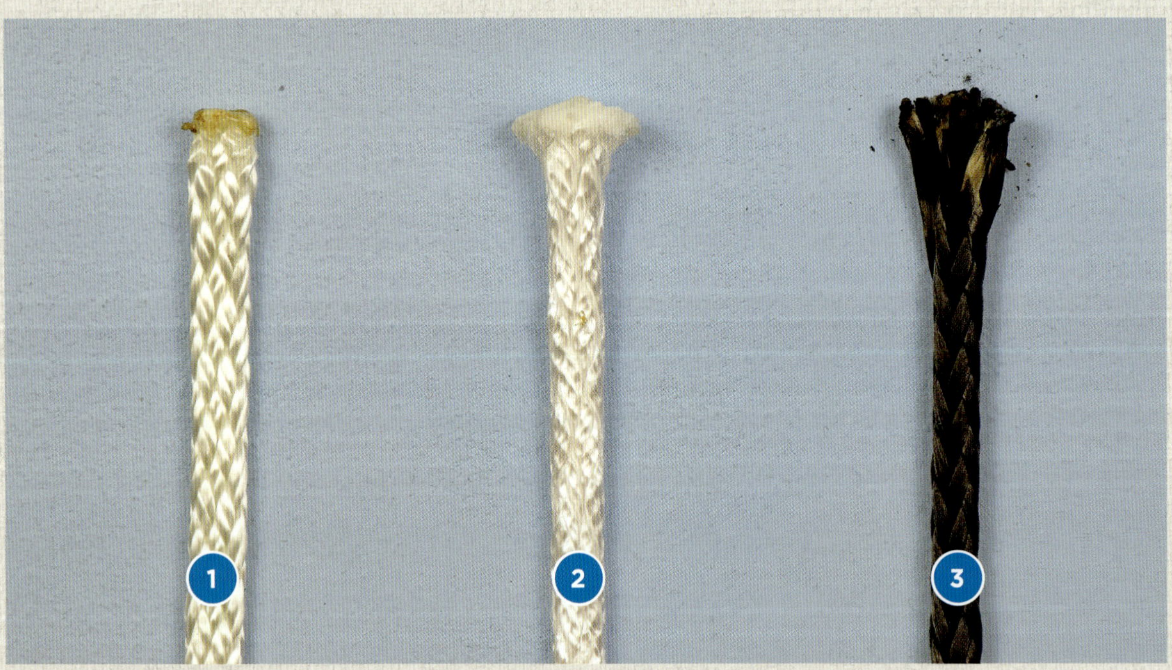

Polyester 1 melts and shows dark edges much the same way as polyamide and polypropylene. Dyneema 2 readily expands into a mushroom shape under heat. Vectran, Kevlar, und Zylon are usually of gold-yellow color but can also be dyed, like this black sample of Vectran 3. These fibers smolder and carbonize in an open flame, giving off small charred particles.

TECHNICAL TERMS

Working stretch is the elongation of a rope under maximum working load. The working stretch is the best indicator to judge stretch in actual use of a rope. Working stretch is measured as a percentage of the length of a rope under tension. Example: A rope of 30-meter length is tensioned with maximum working load and stretches 30 centimeters. This gives a working stretch of 1%. If the same rope is only 5 meters long and tensioned with maximum working load, it will stretch 5 centimeters (1% of 5 m = 5 cm).

Working load, also called safe working load (SWL), gives a limit of load the rope can take repeatedly or permanently without any damage to the rope or risk of breakage. Working load is given as a fraction of breaking strength. Depending on the rope's usage, safety factors need to be applied accordingly. With suspended objects under which people may pass, a higher safety factor is needed (e.g., SWL = BRL), while a halyard or control line on a flat-out race boat like in the America's Cup will have hardly any safety factor, in order to save every ounce of weight possible. A reasonable safety factor in amateur sailing and yachting could be 1/5 of breaking strength, respectively breaking load (BRL).

Eye is a term describing a closed loop in a line. An eye can be knotted like in a bowline, or it can be formed by merely crossing the working end over the standing end while tying a knot. An eye can also be formed by a splice, which is then called an eye splice, irrespective of the kind of rope or the size of the loop (that is to say, the eye).

Breaking stretch is the amount a rope stretches at the point of breaking. Breaking stretch is

All eyes: knotted with a butterfly knot 1, formed by crossing the working end over the standing end 2, spliced in an eye splice 3

measured as a percentage of the length of the rope. It relates to properties of a rope beyond the limits of safe working load, where inelastic deformations take place, which should not be reached in actual use of the rope. Breaking stretch is an interesting parameter, but for evaluating stretch characteristics of a rope, the working stretch is more informative.

Breaking strength or breaking load is the strain under which a rope that is slowly stretched between two attachment points breaks in its free length; that is to say, somewhere in the middle and not at the attachment points. Breaking strength is determined on a tensile-testing machine in a laboratory by actual destructive testing. A calculated or theoretic breaking strength, on the other hand, is derived mathematically from known material data, section area, and material density in the section area of a rope. The actual breaking strength of a rope in use on board a yacht or in other applications is lower than the tensile-testing result of a brand-new rope under laboratory conditions. Wear and aging as well as dirt and particles that found their way into the rope all reduce breaking strength. And not to forget the attachment points, which often form the weakest link, especially if the rope is knotted and not spliced. Breaking strength is often stated in daN (deka-Newton), which correspomds closely to kilogram. For greater loads, kN (kilo-Newton) is used, which approximately equals one-tenth of one metric ton. For an exact conversion, daN needs to be divided by 0.981 (0.1 x gravitational acceleration of 9.81 m/s2) in order to give kilograms, thus stating the breaking strength not as a force, which is physically correct, but rather as a weight, which may be more easily imagined.

Dyneema See HMPE.

End-to-end splice is a splice connecting the ends of one and the same rope to form a closed loop. However, the same technique can also be used to join the ends of two double-braided ropes; compare joining rope splice.

HMPE stands for high-modulus polyethylene, a fiber best known by its trade names Dyneema and Spectra, and is also called UHMPE, meaning ultra-high-molecular-weight polyethylene. This fiber has extreme tensile strength with minimal stretch and is very durable, is resistant to UV radiation, and does not absorb water. These attributes make HMPE an ideal choice for high-performance ropes in marine sports. There are many subtypes of Dyneema used in ropes, with Dyneema SK78 being the most common. Dyneema SK90 and especially SK99 have even -higher tensile strength and are used where high-end performance is demanded. Under constant load, Dyneema deforms and exhibits nonelastic elongation called creep. There are subtypes of Dyneema without creep, such as Dyneema DM20 and others.

Staple yarn or spun yarn is a short length of fiber often produced from polyester. In ropes, staple yarns give a fluffy, soft surface with good grip, which can be advantageous for sheets and control lines that are mainly held and adjusted by hand. Compared to continous yarns, called filament yarns, staple yarns are less abrasion resistant and collect dirt more easily. When splicing it may be a little bit more difficult to pull out single yarns for tapers.

Thimble is a reinforcment usually made of metal that can be inserted into an eye splice. Such a splice is then called a thimble splice. A thimble protects the rope from abrading on the inside of the eye, especially if the attachment point has a rough surface. It also protects the rope from bending too tightly if the rope is connected to a thin-walled object or small-radius bolt or shackle. In the case of single-braided Dyneema or Spectra ropes, it is acceptable to directly connect to a bolt or round object with a radius of 0.6 times the diameter of the rope. In other words, a bolt or fitting with a thickness of 1.2 times the rope's diameter can be attached without a thimble (e.g., a 5 mm single-braided Dyneema line can be attached to a 6 mm bolt, 5 x 1.2 = 6). In high-modulus fiber ropes, only closed or reinforced thimbles should be used. Reinforced thimbles have a bridge or strut welded in. Regular thimbles of the slitted type can deform under

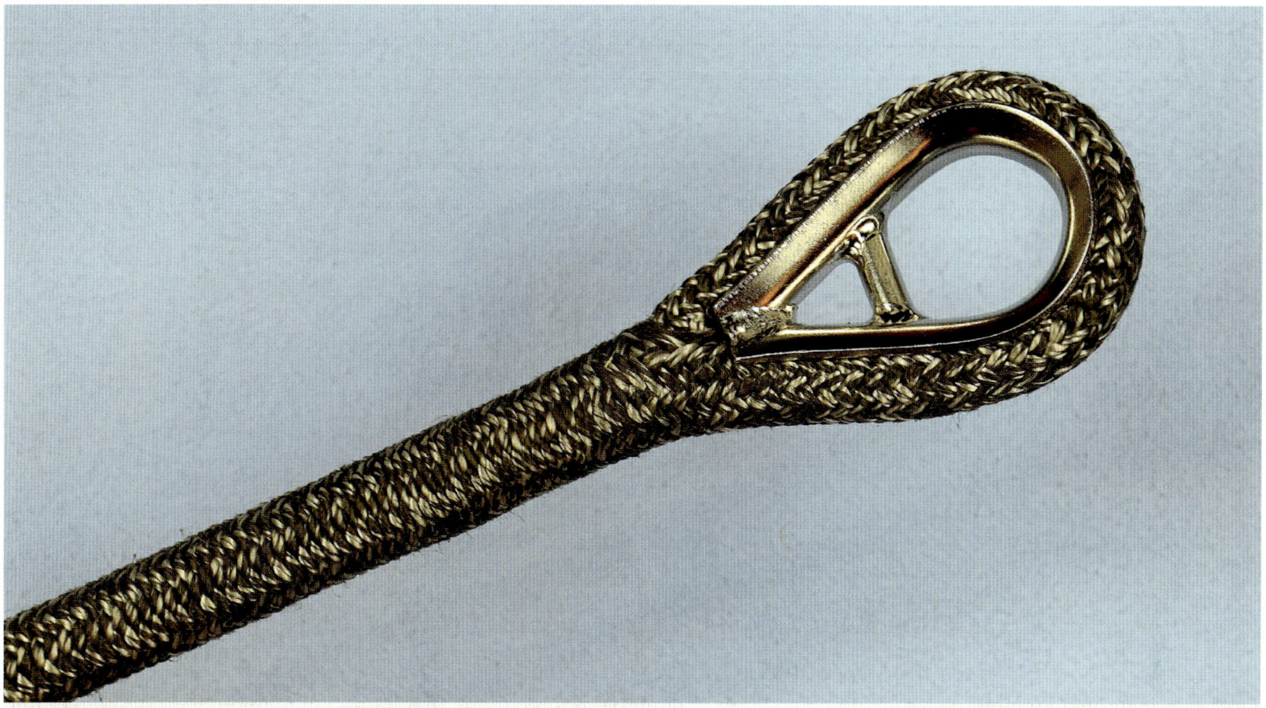

Thimble splice with reinforced thimble for high-performance ropes.

high load. Circular thimbles or high-load eyes from aluminum are often used in lieu of blocks and serve to deflect lines or adjust sheeting angles, especially in freely suspended, so-called 3-D hauling points for foresail sheets. A thimble is not attached to an eye splice, other than it is closely encircled by the eye.

Creep is a nonelastic deformation that occurs under continous static load, especially in HMPE ropes. When the load is released, the rope does not recover to its previous length but remains elongated. This effect prohibits the use of common HMPE ropes in dimensionally stable applications such as standing rigging on a sailing yacht. Creep occurs because under static load, the extremely long molecules in HMPE fibers start to "slide" in longitudinal direction. There are also HMPE types that exhibit no or next to no creep.

LCP stands for liquid crystal polymer, a high-modulus fiber derived from a modified polyester best known by its trade name of Vectran. LCP is very heat resistant and has very high tensile strength, minimal stretch, and no creep but is UV sensitive.

PBO stands for polybenzoxazole or poly(p-phenylene-2.6-benzobisoxazole) and is best known under its trade name of Zylon. PBO is a high-modulus fiber of the highest strength and has the least stretch without any creep. This fiber is very resistant to heat but is not durable if exposed to sunlight or any UV radiation. PBO is also sensitive to moisture and continual bending. In sailing, PBO is mainly used for composite standing rigging on race boats.

Stretch describes the elongation of a rope under tension. It can be divided into an elastic and inelastic portion of stretch. Rope with very little stretch will show an elongation of less than 2% of its length under working load, which would be referred to as 2% working stretch.

Vectran See LCP.

Joining rope splice is a splice that connects one rope with another; compare with end-to-end splice.

Zylon See PBO.

SPLICING TOOLS FOR BRAIDED ROPE

There are quite a number of different splicing tools available, but it is absolutely not necessary to acquire all kinds of variations in order to splice all kinds of braided ropes. Basically, splicing tools fall into two catagories: tools to push the end of a core or cover inside a rope, and tools to pull those ends inside. Tools to push inside can be used for simple splicing tasks such as loosely braided ropes or hollow braids. Tools to pull the ends inside can be used for all splicing tasks, as well as for very tightly braided ropes, and make work generally much easier.

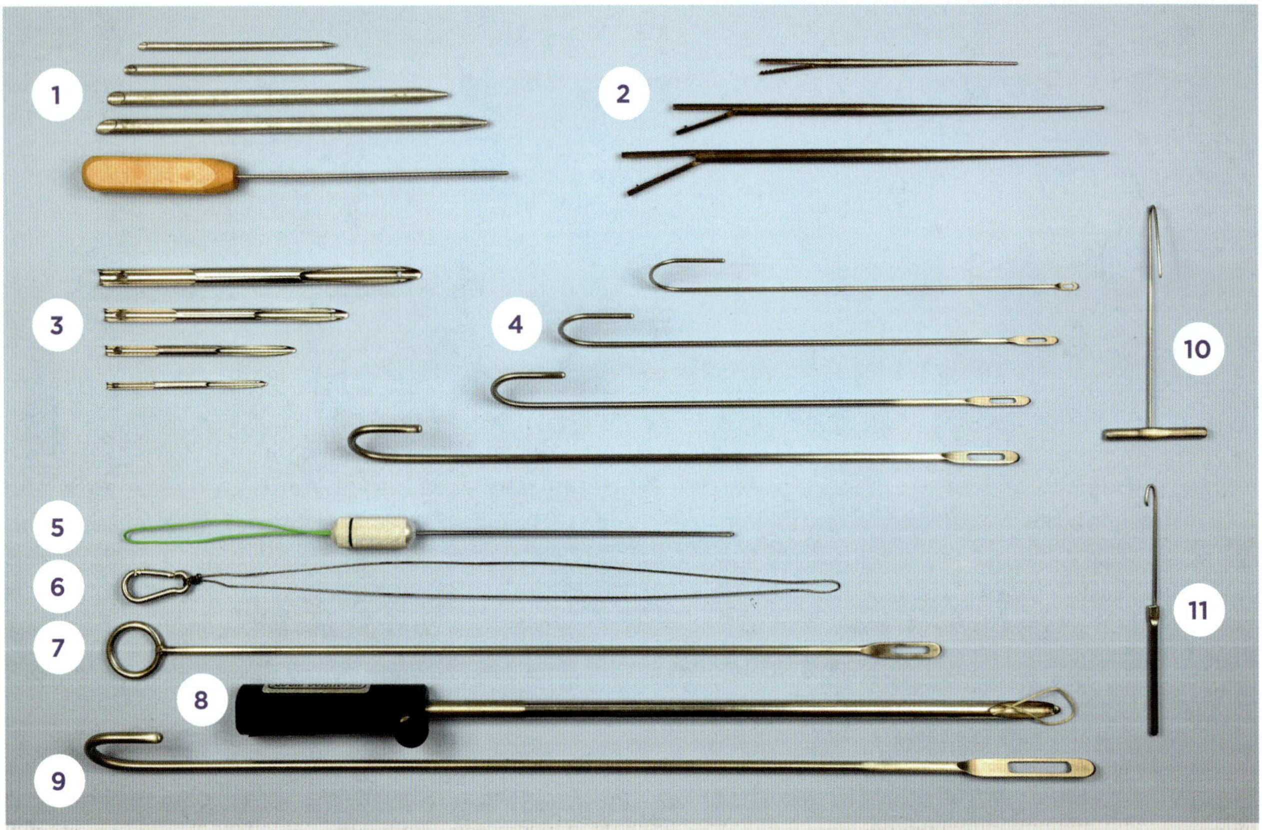

Selection of different splicing tools

TOOLS TO PUSH INSIDE

This category of splicing tools is suited for single- and double-braided ropes that are relatively loosely braided or plaited.

❶ Fids and pusher, classic splicing tools for loosely braided ropes. Made of aluminum by Samson Ropes and others.

❷ Fids with opening ends to attach a core or cover. For loosely braided ropes. Made of steel by Liros.

❸ Selma-Fids, hollow fids made of stainless steel. The end of a core or cover is placed inside and is held by a small hook. These tools are produced by Selma and are best used for hollow braids.

TOOLS TO PULL INSIDE

This category of splicing tools is suited for all kinds of single- and double-braided ropes.

❹ Splicing needles for all sorts of splicing tasks, including very tight and stiff double -braided ropes. Made of stainless steel, forged and machined, available from EMF Marine and Marlow.

❺ Splicing tool consisting of a length of bent wire attached to a handle. Suited for small diameters, made of stainless steel and aluminim by D-Splicer.

❻ Splicing tool, self-made of bent wire. With this tool, great pulling force can be applied if high-strength wire such as in piano strings is used and attached firmly to a shackle or carabiner. Great pulling force is needed only for specialized ropes sometimes used by industrial climbers and arborists.

❼ Splicing needle with a welded attachment point in the form of a ring, to apply greater pulling force for specialized hollow braids. Stainless steel, forged and machined, available from EMF Marine and Marlow.

❽ Splicing needle or wand with a small sling to hold the end of a cover or core. The sling can be drawn close by sliding back the handle. Stainless-steel tube with brass rod inside, available from Brion Toss Riggers.

❾ Large splicing needle for rope diameters up to 36 mm or 1 inch. Stainless steel, forged and machined by EMF Marine.

❿ Hook, self-made from stainless-steel wire to pull a bight of single-braid rope back through a capsized braid when making the self-locking brummel splice without having to pull the standing end through (see page 51). Suited for rope diameters of 10–20 mm.

⓫ Easy-Hook, a simple stainless-steel hook on a handle as above, but for rope diameters of 4–8 mm, available from EMF Marine.

What else is needed? A measuring stick, a sharp knife (it's best to use a knife with a curved blade, not serrated; no box cutter or carpet knife), a cutting board, scissors, sewing needles (preferably use sailmaker's quality).

BASIC SPLICING SKILLS —WHAT TO BE FAMILIAR WITH BEFORE STARTING TO SPLICE

SPLICING SKILL 1

GETTING THE CORE OUT OF THE COVER VIDEO SS 1

With all double-braided ropes (kernmantle ropes), it is possible to extract the core at any given point away from the ends of the rope. This is because core and cover are not affixed to each other*. The core merely lies inside the cover. Load is transfered onto the core only by friction and only when the rope is under tension.

The degree of difficulty when extracting the core largely depends on how tightly the cover has been braided. Stiffer ropes demand more skill, but the principle always remains the same. A tightly woven cover with a 1-over-1 braid in a high-modulus, mixed-fiber rope makes work somewhat harder. But do not dispair—skill comes with practice, and quickly.

Here is an easy task to start with: 12 mm, double-braided rope with 20-plaited, loosely braided cover.

Use a metal needle or a small fid with a rounded tip. Prick exactly between the strands of the cover braid into the rope. Insert the needle at an angle toward one side. Make sure to insert the needle between whole strands, which consist in this case of pairs of yarns. Do not insert the needle between individual yarns or even pierce it through the middle of a yarn.

Wrong! Here the needle is inserted between the two individual yarns of a strand.

Stretch the cover braid to one side, using the needle like a lever.

Prick into the same opening but now toward the other side. Again stretch the cover braid, using the needle to apply leverage. Repeat stretching the cover braid alternately to both sides until an opening is created and the core becomes visible.

Bend the rope at the opening. Pierce into the core carefully from the side. Be very careful not to pierce through or under any of the cover yarns. Lift the core up and simultaneously push the cover to the sides, using both of your thumbs.

This is what it should look like when pulling out the core. Try to pull the core out as a whole, undamged. Do not rush to get the core out; take your time and do not use excessive force. From the side that leads to the short end of the rope, the core can then be pulled out completely.

A more difficult task: thin-diameter rope, with very tightly braided, 24-plaited cover consisting of a mix of high-modulus fibers, in this case Dyneema and Technora.

Bulge the rope several times at the point where the core is to be extracted, by holding the rope as shown and strongly pushing together the short section between your hands.

Pierce between whole strands into the cover braid. Using a small fid or needle with a well-rounded tip makes it easier to insert the tool between the strands and not pierce through individual yarns.

Push the fid as far as possible under the cover, toward one side. There shouldn't be any soft cracking sounds. If there are, then some fibers ruptured, indicating that you did not perfectly hit the spot between the strands.

Strech the cover braid to one side. At first it seems like the streching has only little effect.

Now push the fid under the cover, but toward the other side. Stretch the cover repeatedly, using the fid for leverage. Step by step, a small opening will appear in the cover braid. The emphasis is on "step by step." The braid can only slowly yield to the stretching. It is not possible to create an opening by stretching the braid just one or two times. It must be stretched repeatedly and alternately to both sides.

Here the game is already up for the core. As soon as a little opening like this is created, where the core becomes visible, then it also can be gotten out. Stretch only a few more times to both sides and then …

... prick the fid or needle under the core. Push the cover yarns to the sides, using both of your thumbs, and carefully lift up the core until it can finally be pulled out completely.

Mission accomplished! The core is pulled out as a whole, and it is undamaged. Take the core with your fingers and pull at the side that leads toward the short end of the rope to extract it completely.

SPLICING SKILL 2

WORKING WITH THE SPLICING NEEDLE

VIDEO SS 2

Splicing needles have the great advantage over fids that the ends of the core or cover can be pulled into the rope with great force, as opposed to having to be pushed inside with a fid. That makes work a lot easier, especially when splicing tightly braided, stiff ropes. A further advantage is that the ends of the core or cover can be attached to a splicing needle without creating any increase in diameter.

INSERT the splicing needle exactly between the strands of the cover braid. With very tightly braided covers, a small opening can be created by first using a small fid or spike.

TIP: When splicing double-braided ropes, insert the splicing needle on the opposite side of where the core has been extracted.

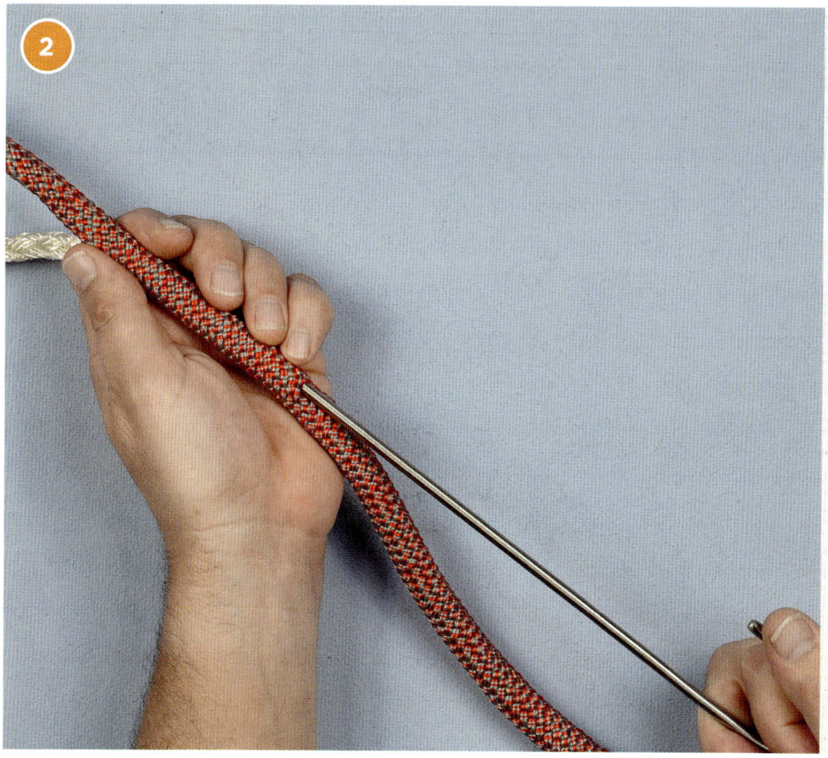

PUSHING THROUGH Hold the rope with a loose grip. Do not squeeze the rope in your hand. Push the splicing needle along, going exactly between core and cover. The splicing needle must not go inside the core or under any of the core yarns!

TIP: If unsure whether you have accidentally inserted the needle into the core, pull out more length of the core. That also creates more space, almost in the form of a little tunnel, between core and cover, and the needle can be pushed through more easily.

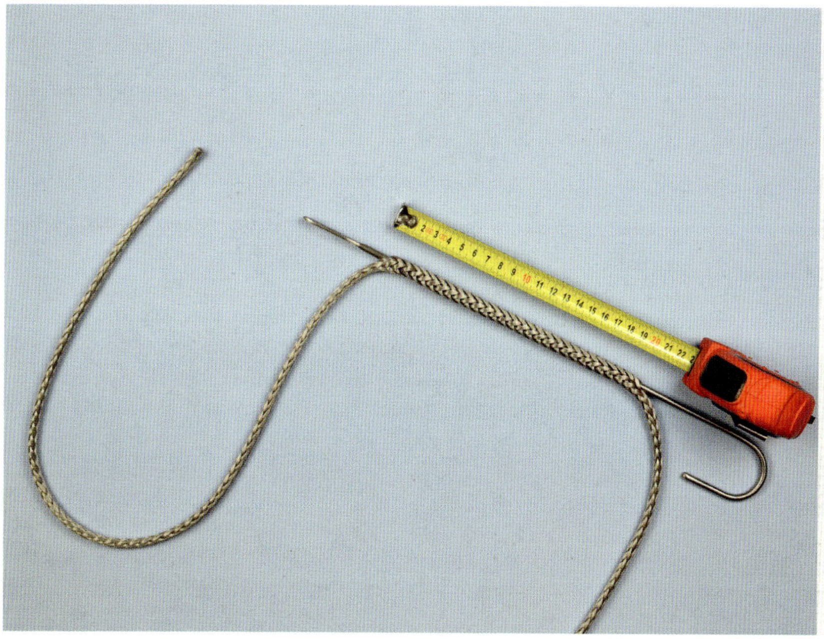

With some splicing work, especially when splicing single-braided Dyneema ropes, there is the additional advantage of using splicing needles that the desired length of the spliced section can be measured before actually pulling the rope's end inside. When the splicing needle is pushed along the inside of the rope, the rope thickens and is shortened at the same time, to very much the same length as the finished splice will have.

EXITING Exit with the splicing needle again, exactly between the strands of the cover braid. Widen the exit opening slightly by stretching it with the splicing needle, so that it is easier for the core to be pulled inside during the next steps of splicing.

***TIP:** When splicing double-braided ropes, exit with the splicing needle on the same side where the core has been extracted. That way, any torsion of the rope within the section of the eye splice is avoided.*

CHECKING Before continuing any further, it is very important to check whether at some spot the splicing needle has been accidentally inserted into the core. Even if only a few fibers or yarns of the core got on the wrong side of the needle, it will be very hard if not outright impossible to finish the splice. To check for this, pull out more length of the core. The core must easily slide along the splicing needle. If there is any resistance, pull out the splicing needle and some length of the core and then insert the splicing needle again. Accidentally pushing the splicing needle under only one or two yarns of the core is the most common mistake when splicing double-braided ropes.

ATTACHING to the splicing needle. When splicing tightly braided ropes, open up the braid at the end of the core and thread only about half of the core yarns onto the needle. The other yarns are cut off carefully. They should not stick out to the side. Thus any increase in diameter at the attachment point with the splicing needle is avoided, which is prerequisite when splicing tightly braided ropes.

INFO: With loosely braided ropes and with single-braided ropes like Dyneema or Spectra 12-strand ropes, the end can be threaded onto the splicing needle as a whole. Push the end through the eyelet of the splicing needle for only a few millimeters, bend it slightly so that it won't fall out, and pull it inside the rope.

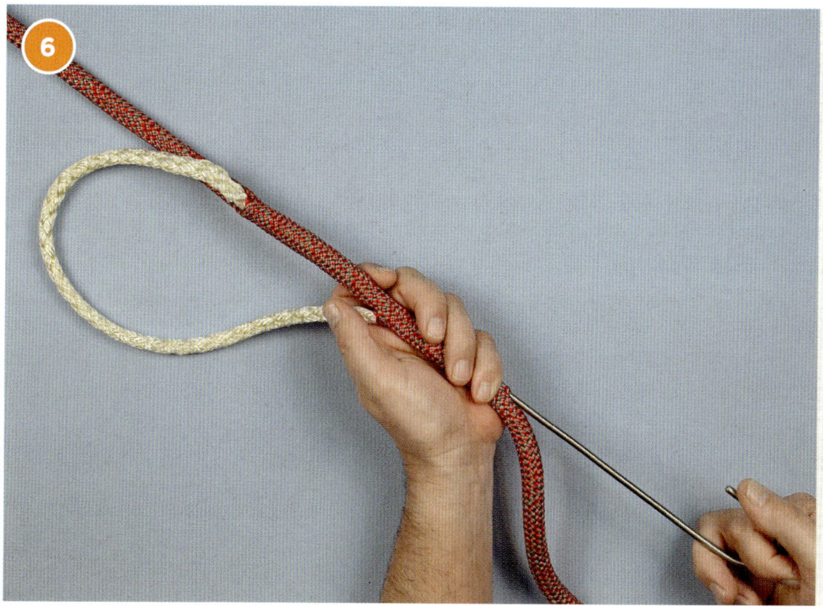

PULLING THROUGH Hold the rope away from the point where the core is pulled inside. Hold it with a loose grip; do not squeeze the rope. The cover must be allowed to bulge. Slightly twisting the splicing needle facilitates pulling the core inside. Milk back the cover repeatedly while pulling the needle through, until the core emerges again at the other end.

SPLICING SKILL 3

TAPERING

VIDEO SS 3

Taperig is a technique to steadily reduce the diameter of a braid over a certain distance. To achieve this, single yarns or whole strands are being pulled out of the braid at regular intervals and then cut off.

Tapering is an essential technique in order to achieve maximum breaking strength in the finished splice. Any uneven change in diameter, which would occur without tapering, creates a weak spot. The larger the rope diameter, the more important tapering becomes with regard to breaking strength. Only with rope diameters below 4 mm are the effects minimal.

But there is also another equally important reason for tapering. With carefully tapered cover and core ends, it is a lot easier to close an eye splice in double-braided rope when you have to milk back in the final stages of the work. The tapered ends are practically pointed and can be drawn into the rope so much easier than had they been left blunt and untapered.

With the rounded tip of a spike or small fid, lift up a single yarn. Then pull it with your fingers at the side that leads to the short end of the rope. Pull the yarn out completely. Do not pull on the side that leads toward the long end, the standing end of the rope, in order to avoid distorting or creating uneven tension in the braid.

TIP: *Just slightly lift up a single yarn with the spike. Retreat the spike for a moment and then go under the yarn again with the spike to lift it up completely. That way, the yarn can realign and you avoid accidentally catching any of the fibers of the yarns next to the one you want to pull out.*

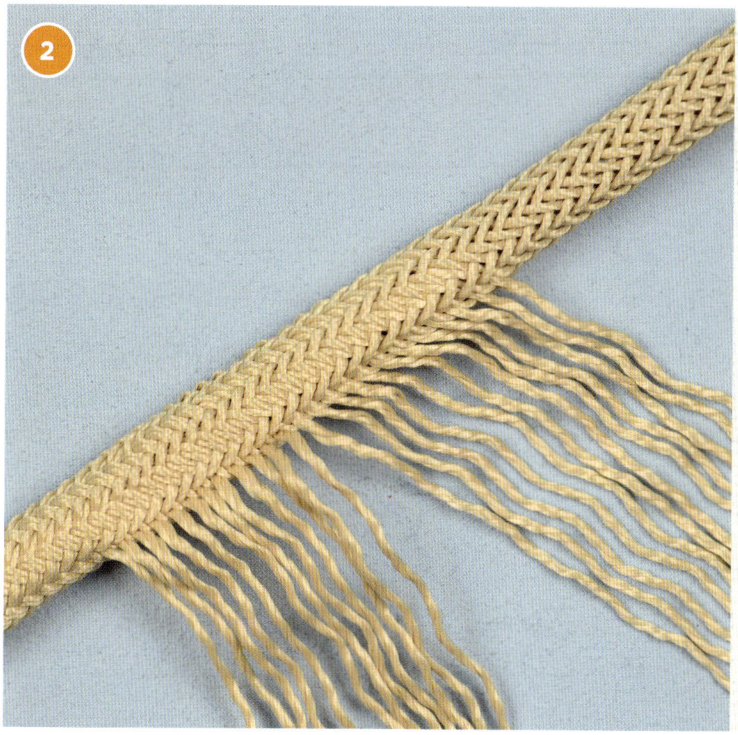

The degree of a taper can be made to suit, but normally about half the yarns of a braid will be pulled out. The most important point is that the taper has an even form and that the diameter decreases as regularly as possible. Different patterns of how to pull out yarns can be found. In the case of this 20-plaited or 20-strand cover braid, first one single yarn was pulled out of all the strands that go to the left, then one single yarn was pulled out of all the strands that go to the right. This is a simple and easy-to-follow method. At the end of the taper, all strands consist of only one yarn, while at the beginning of the taper, each strand has two yarns. That way the braid has been halfed.

A slightly different method possibly gives the ultimate in regularity. Here it was always alternated between pulling out a single yarn that goes to the right and a single yarn that goes to the left. The equal distances between the spots where the yarns are pulled out can be chosen according to the desired length of the taper. In this case, every second strand was used to pull out single yarns, but it could also have been every third or every fourth strand in order to spread the taper over a greater length. Comparing the braid at the beginning and at the end of the taper again shows how the braid has been halfed. At the end, every strand consists of a single yarn, while at the beginning every strand has two yarns.

TIP: *Incidentally, observe how the single yarns are twisted. In good-quality ropes, the yarns that go to the right should be twisted to the left—that is, having an S lay, and the yarns going to the left should be twisted to the right, having a Z lay.*

Tapering of a 12-strand Dyneema or Spectra rope. Six strands have been pulled out at regular intervals, alternating between strands running right and running left. The braid has been halfed but has not been compromised or dissolved.

At the very end, unlay the braid and cut the end at a sharp angle to finish the taper.

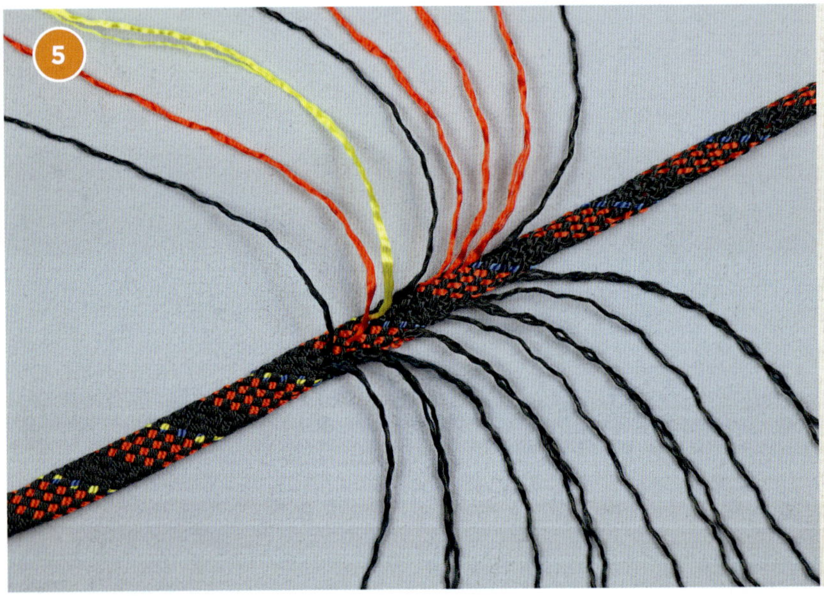

With some braids, the strands that go to the right and the strands that go to the left consist of differing numbers of yarns. This 16-strand, one-over-one cover braid has three yarns in all of the strands that run to the left, but only two yarns in all of the strands that run to the right. To effect a taper over a relativly short distance, yarns can be pulled out in the following order, as shown:

- From the strands that go to the right, only one yarn is pulled out of every second strand. (These are laid out pointing upward in the picture.)
- From the strands that go to the left, yarns are also pulled out of every second strand, but alternating from strand to strand. One time one yarn is pulled out, then two yarns in the next strand, then again one yarn, and so on. (These yarns are laid out pointing downward in the picture.)

When tapering 24-strand or 32-strand, one-over-one braids, it is also acceptable to pull out whole strands instead of single yarns.
For example, from the strands that go to the left, every second strand could be pulled out as a whole, and the same with the strands that go to the right.

TIP: ***Many different patterns can be found to pull out yarns. The important point is to achieve regularity and evenness as well as possible. On the other hand, it won't be the end of the world if one or two yarns too many are pulled out at some spot. Just try to make the taper nice and even.***

After the yarns have been pulled out, they are cut off flush.

SPLICING SKILL 4

MILKING BACK

VIDEO SS 4

All splices in double-braided ropes (kernmantle ropes) are closed and brought into their final size and shape by milking back the bulged cover braid. For milking back effectively, it is necessary to attach the rope to a strong point with the initially made loop knot, so that considerable pulling force can be applied. But it is not just about sheer strength; even more important is the correct technique and some practice when milking back.

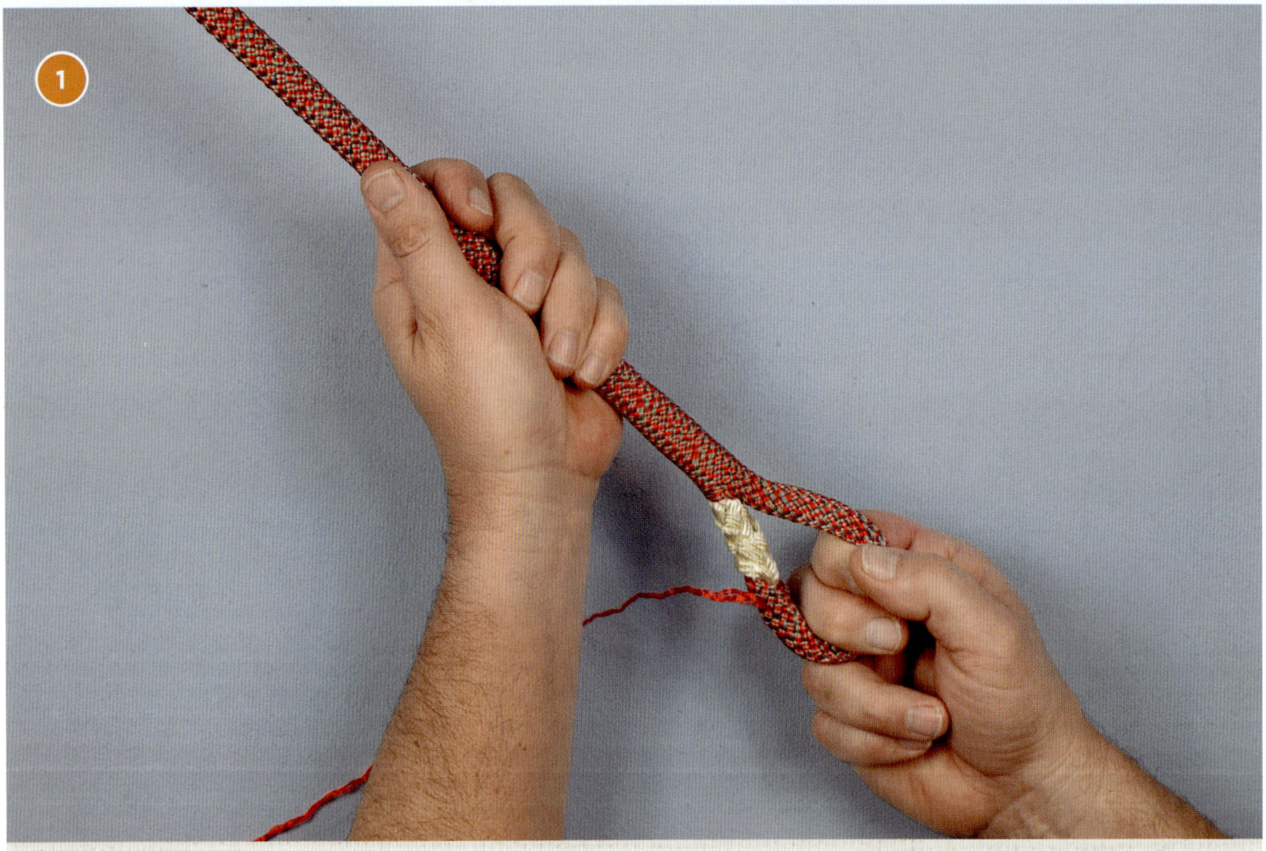

Up to this point, milking back is fairly easy. It is the last inch or two until the cover marker yarn reaches the entry point when more strength needs to be applied. If it›s too easy to close the splice up to the cover marker yarn, then that would indicate that a mistake has occured, most likely when measuring the so called core shortening. Always stick to the following important rules to milk back and close the splice successfully:

- Push all of the bulged cover braid toward the splice, starting all the way back at the loop knot.
- Always hold the entire length from the loop knot to the eye of the splice under tension. That means pulling at the eye with one hand and milking back with the other. At no point should the rope at the eye be slackened when closing the splice!
- Try not to pull just statically. Hold the eye with one hand; the other hand holds the bulged cover braid and jerks the line dynamically, using body motion to bring the line taut with a snap.

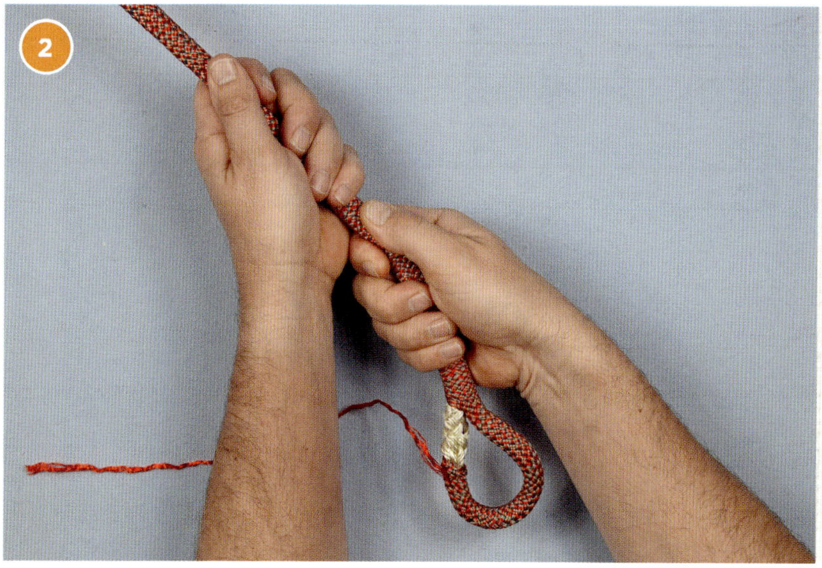

Wrong! Milking back this way, the eye is slackened and the rope at the eye will bulge up.

This happens when the eye is not held under tension while milking back. The splice does not close completely, and the cover around the eye is bulged up. Around the eye, the rope is thicker than in the rest of the rope's length, and that should not be.

It is easy to correct a bulged cover inside the eye. Milk back firmly and repeatedly from the cover marker yarn around the eye circumference and down the splice.

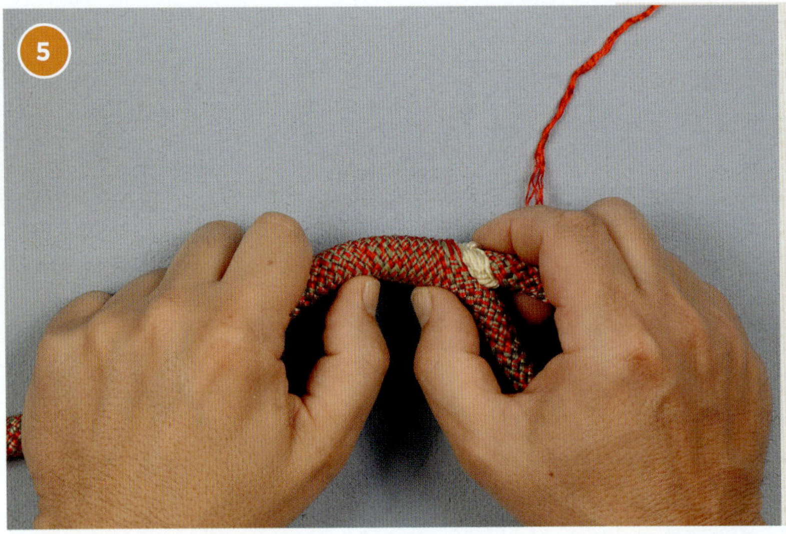

If it is hard to close that last bit: Bend the splice back and forth several times to make it more pliable. This is better and more effective than pounding the splice with a hammer or mallet. Milking back and closing the splice normally takes more than just one pull. Milk back, release tension, bend back and forth, apply tension, milk back, and continue until the splice closes completely.

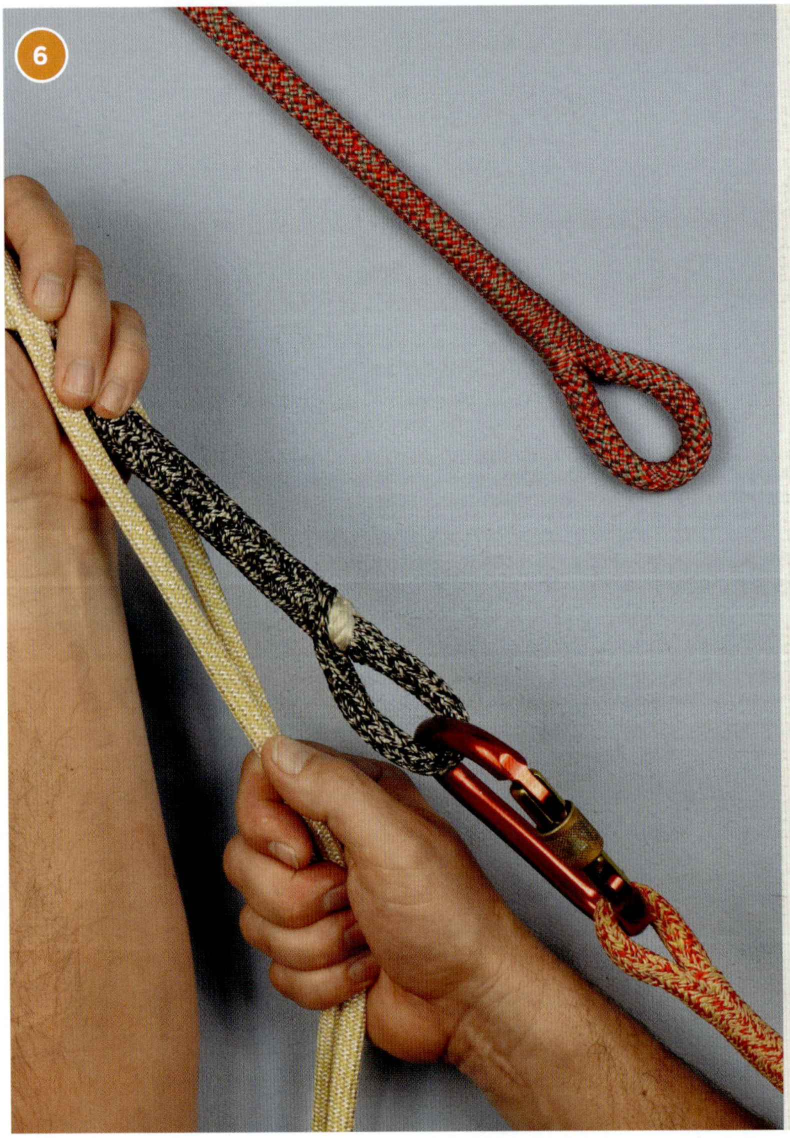

Ropes with high-modulus fiber cores such as Dyneema or Spectra often call for more pulling strength when milking back. Tension the line between loop knot and eye splice, using a purchase (pulley, tackle) or a winch. Apply several hundred or a thousand pounds of pulling force. The core sometimes gets bunched up a little right before where it is drawn inside. But this will completely disappear when the splice is closed under sufficient tension. Milk back over the tensioned rope. You may want to wear work or sailor's gloves to avoid getting blisters. The method shown in the picture uses a short length of hollow cover braid that is wrapped around the rope. This is very effective but also needs some practice. The right hand pulls and effects pressure on the rope; the left hand pushes the wrapping turns and thus the bulged cover over the splice.

SPLICING SINGLE- BRAIDED ROPES (ROPES WITHOUT A CORE)

1.1 EYE/THIMBLE SPLICE IN CONVENTIONAL FIBER AND MIXED-FIBER ROPES

Single-braided ropes made of polyester (PES) or polyamide (PA) or polypropylene (PP), also in combination with high-modulus fibers.

Ropes of single-braided construction can be very soft and pliable. It gets especially interesting if single-braided ropes are manufactured using two different materials, one for good grip like PES or PA and one for high strength and virtually no stretch like HMPE, Dyneema, or Spectra. In this case, the positive characteristics of those two materials are not split into a core and a cover braid of a rope but are, rather, braided parallel to each other into a single braid. Modern ropes of this composition are called mixed-fiber ropes. They are very flexible, create very little friction in blocks, and are ideal for sheets and control lines that are held and adjusted by hand. Some of these ropes have a small-diameter string in their middle. This is not a core and has nothing to do with the strength of the rope; it merely serves to give the braid a nice, round section. This string needs to be removed over the full length of the splice.

START

VIDEO 1.1

Construction:	single braided (no core)
Material:	mixed fiber, Dyneema, Spectra combined with Polyester, etc.
Difficulty:	easy (if the end is tapered carefully)
Splicing tools:	splicing needle or similar pull-through tool

1

Measure the end that will be pulled inside. The minimal length needs to be 24 times the diameter of the rope. Here, 24 x 8 mm diameter = 192 mm or roughly 19 cm.

At the measured distance, pull out a single yarn. Use a small needle or spike. It›s best to pull out one of the high-modulus fiber Dyneema yarns, which are the white ones in this case. Pull only on the side of the yarn that leads toward the short end of the rope. Avoid tensioning the yarn toward the standing end, so that the braid does not get uneven or distorted. (The second-best method is not to pull any such marker yarn out but to mark the measured length with a felt pen.)

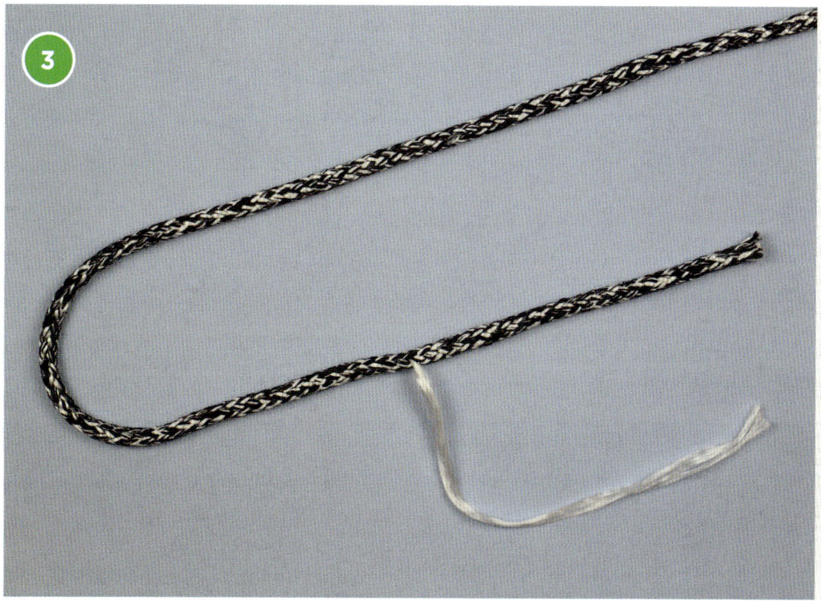

A single yarn has been pulled out and marks the length of the rope's end that needs to be spliced inside. The yarn also serves the purpose of securing the splice by using it later to stitch through the finished splice a couple of times. (If marking the distance with a felt pen instead, the splice will need to be secured with a separate short length of whipping twine by making several stitches or applying a palm-and-needle whipping.)

***TIP:** In case the rope has a small-diameter string in the center of the braid, this string has to be removed at this stage. Measure twice the length of the end to be spliced, plus the eye circumference, plus a shortening of about four times the rope's diameter, and pull the string out from the end of the rope. Cut the string at the total of the added-up lengths. Firmly milk back the bunched rope. The video also shows all the steps of how an inner string is removed.*

4

Taper the end of the rope by about half the yarns. To do so, pull out yarns at regularly staggered intervals. Start the taper about two rope diameters away from the marker yarn. The chapter section "Basic Splicing Skills 3" on page 30 shows all the details of tapering.

5

The pulled-out yarns are cut off flush. Be careful not to accidentally cut the marker yarn off!

6

Now the end has a nice taper. Note that the braid has not been dissolved. The more careful the taper is executed, the easier it is to pull the end inside.

Place the end next to the standing end of the rope and form the eye in the desired size. Insert the splicing needle a few centimeters farther away from opposite the rope's end, to allow for the shortening effect when the end is pulled inside. Push the splicing needle exactly along the center of the braid and exit opposite the marker yarn.

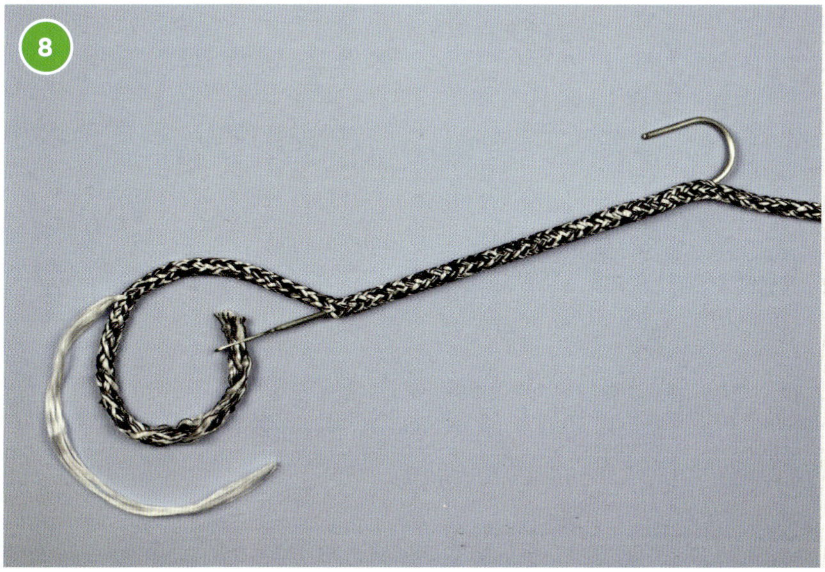

Thread the tapered end onto the splicing needle.

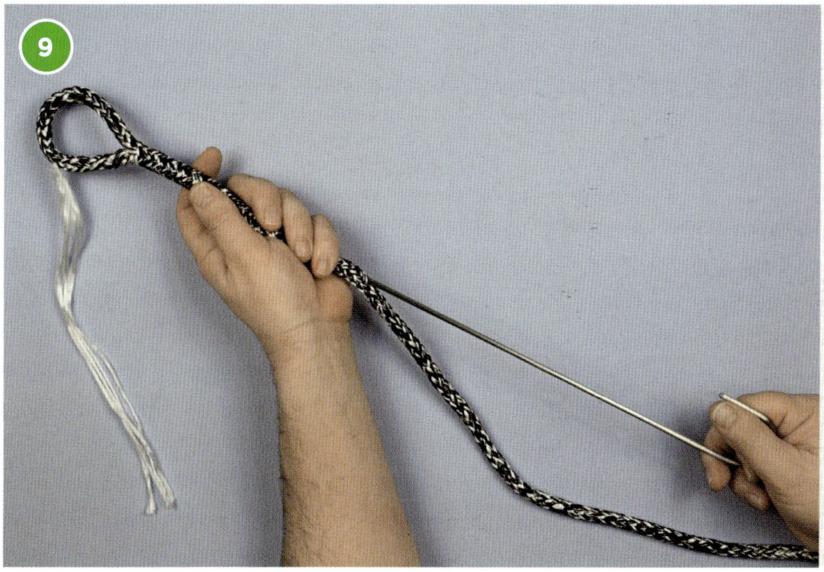

Pull the end inside.

When the tapered end exits, hold it firmly and pull until the marker yarn has reached the entry point.

Cut the end at a sharp angle to finish off the taper.

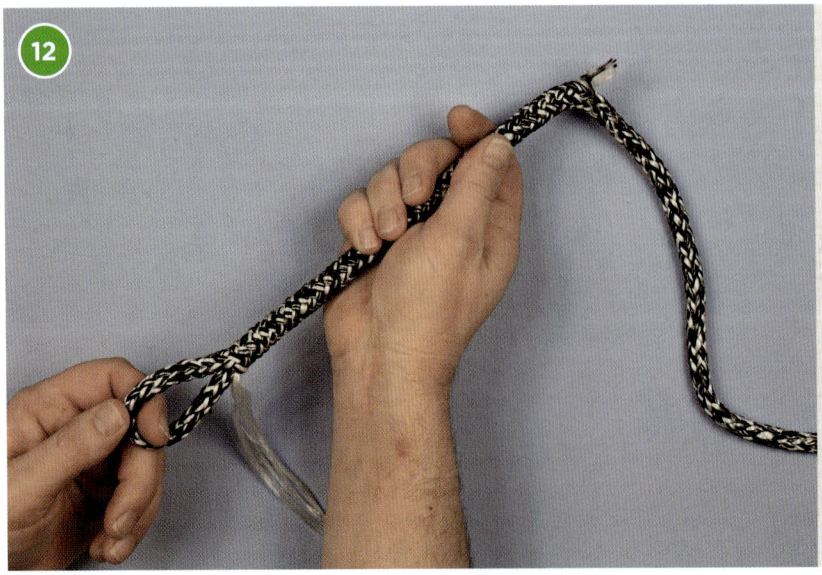

Starting at the marker yarn, milk back firmly until the end disappears inside.

Secure the throat of the splice by stitching the marker yarn through the middle of the rope. Make about four or a few more closely spaced stitches.

Finally, cut the marker yarn off flush.

This is what the finished splice should look like. There are no abrupt changes in diameter visible or tangible. The splice has an even decrease in diameter from the thicker throat down toward the standing end.

Testing for strength: If the rope breaks outside the length of the splice, it is fair to say that the splice has held 100% of the breaking strength of the rope. In the case of this 8 mm mixed-fiber rope made of Dyneema and polyester, the breaking strength was 25.33 kN or 2.58 metric tons (5,688 lbs.). The thin string in the middle of the rope was leading up to the beginning of the splice and was pulled out when the rope broke.

1.2 EYE/THIMBLE SPLICE IN HIGH-MODULUS FIBER ROPES (DYNEEMA, SPECTRA, ETC.)

Single-braided ropes from high-modulus fibers such as Dyneema, Spectra, Vectran, and others. These ropes can also be called hollow braids.

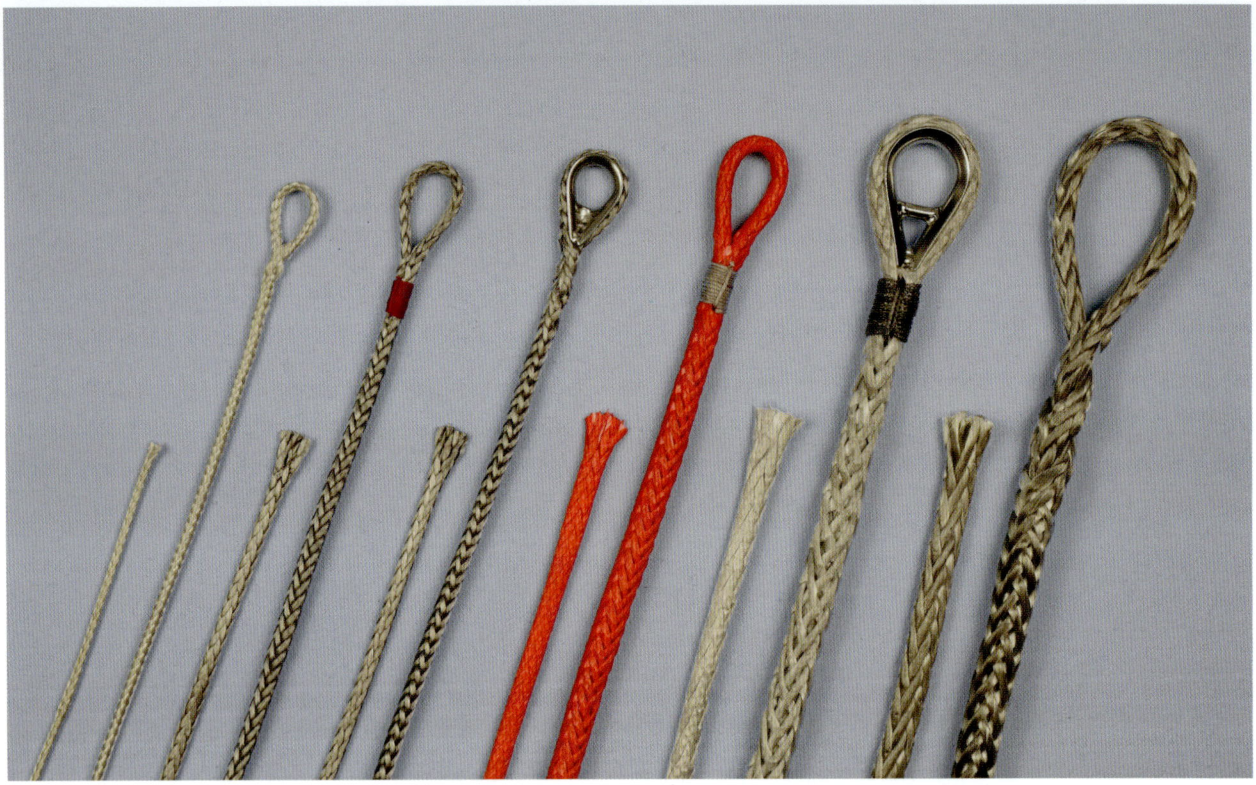

HMPE ropes (Dyneema, Spectra) have a special coating of the yarns that solidifies the surface of the rope and improves abrasion resistance. Additionally, the coating can be pigmented to give the rope any desired color. Dyneema itself is white. Besides the coating, the so-called thermal setting under tension is another very exacting enhancement and decisively determines the rope's quality (see also the chapter "Modern Rope Know-How") .

Given the same diameter, single-braided Dyneema or Spectra ropes are comparable to steel wire, having even superior strength and about equal stretch characteristics. Only in terms of weight is there a great difference: HMPE ropes are about 90% lighter! And they are much easier to splice!

Single-braided HMPE ropes are often used as strops, running backstays or in purchase systems. Because of their slippery surface, they are less suitable to be held or adjusted by hand. Their ends must be attached by splices, because knots decrease their breaking strength by about 65%! The only exception is correctly made lashings (transfilage); see chapter 1.6. Any commen knot, such as a bowline, for example, will just slip and pull open under high load. The same applies for belaying on a cleat; single-braided HMPE ropes can slip even with numerous cross-turns on a cleat. Used in stoppers or on winches, these lines will wear quickly unless an additional cover braid is spliced onto the rope at that point; see chapter 2.6. On the other hand it is very easy and quick to make eye splices with very high breaking strength!

START

VIDEO 1.2

Construction:	single braided (no core)
Material:	high-modulus fibers (e.g., Dyneema)
Difficulty:	easy; only the method with the self-locking brummel splice, where the standing end is not pulled all the way through, demands a little more practice.
Splicing tools:	splicing needle or fid

Method without brummel splice—here the end is simply spliced into the standing end and secured.

1

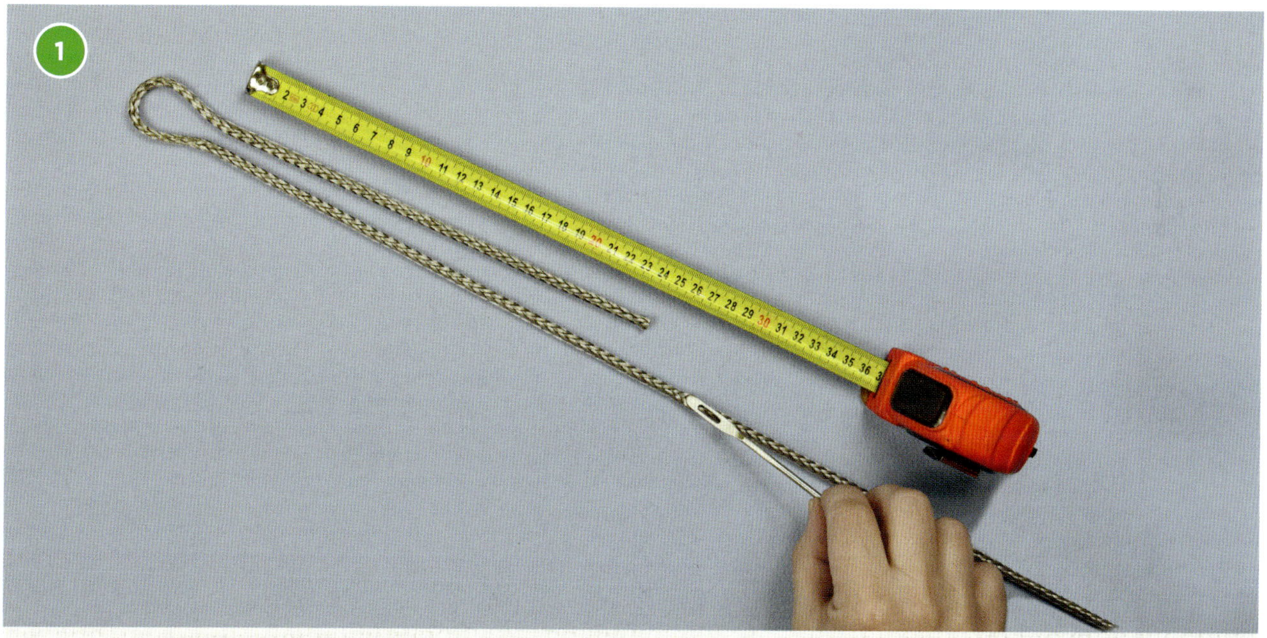

Measure the splicing length. With HMPE ropes, this is 48 times the rope's diameter. Here, 48 x 5 mm diameter = 240 mm or 24 cm. (You can round this to 50 times the rope's diameter, which makes calculating easier.)

Form the eye in the desired size.

In order to pull the whole splicing length inside, it is necessary to insert the splicing needle over a slightly greater length than 24 cm. This is because the rope not only will increase in diameter over the spliced section but will also shorten when the end is spliced inside! So, insert the splicing needle not exactly opposite the rope's end, but a little bit farther down the standing end, just like shown in the picture.

You may want to mark the beginning and end of the eye circumference with a felt pen, although with more practice this is not really necessary.

TIP: *Sometimes the splicing length is recommended to be 100 times the rope's diameter. This includes a very large safety margin that goes beyond the safe working load and is hardly necessary in ropes of less than 10 mm diameter. The required splicing length for maximum strength is proportionally greater in large diameters. With common rope sizes, 48 or 50 times the diameter is fine. Only if absolute maximum strength is required and the rope's coating is especially slippery should the splicing length be increased to up to 100 times the diameter.*

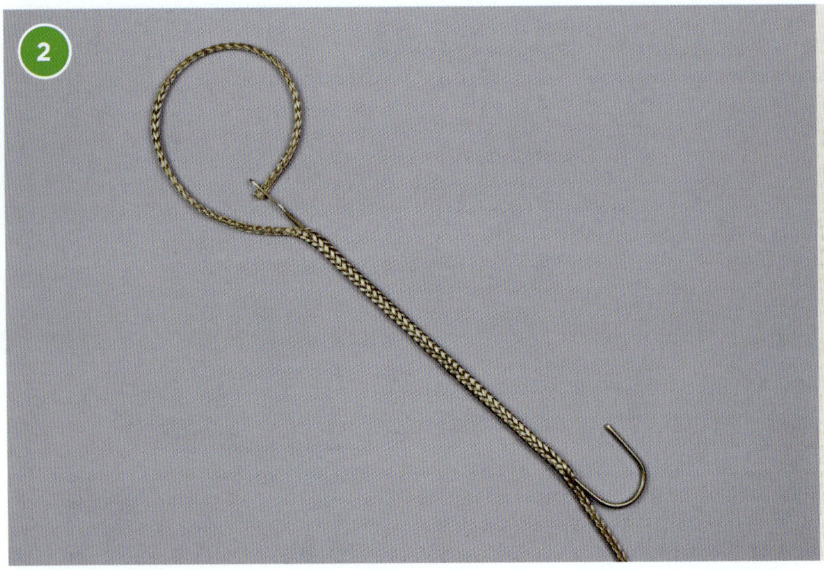

Push the splicing needle through, up to the beginning of the eye circumference. Exit at this point with the splicing needle and thread the rope's end onto the needle.

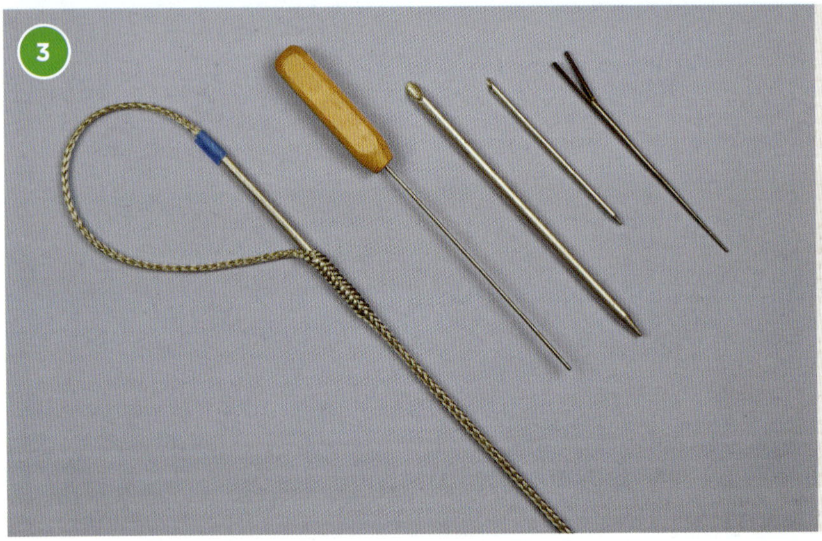

If working with a fid (push-through tool), you have to insert the fid at the point where you would exit with the splicing needle (pull-through tool), which is at the beginning of the eye circumference.

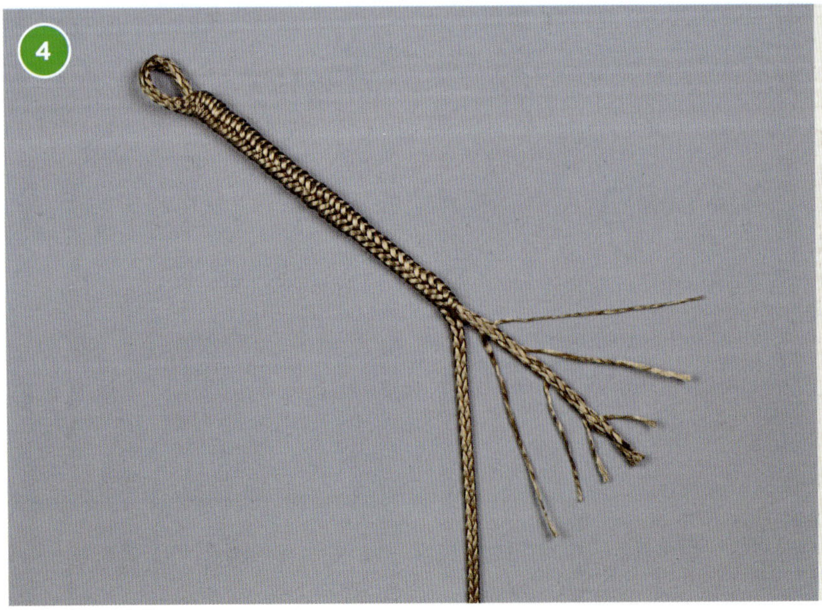

Pull the end completely through and taper the end over the last third of its length, by pulling out single strands in regularly staggered intervals until the braid is halfed. That means half the strands the rope is made of needs to be pulled out. Pictured here is a 12-strand Dyneema rope, so six strands have been pulled out. Further details on tapering can be found in the chapter "Basic Splicing Skills" on page 20.

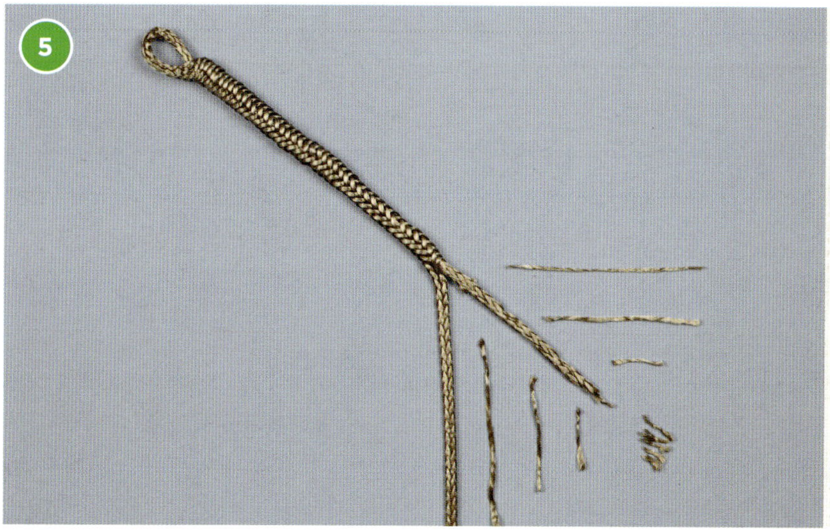

Cut the strands off flush and also cut the very end of the rope at a sharp angle so it terminates to a point.

Adjust the eye to exactly the desired size and repeatedly milk back firmly over the standing end. The rope's end disappears inside.

The splice is finished, but it needs to be secured so that the end cannot slip or pull out under any circumstances, especially under recurring low loads. One method to secure the splice is to loop it through in the way the block is attached to the rope in the picture. By looping through, the splice is reliably secured, and this method has the added advantage that the connection can also be loosened again.

Otherwise the splice needs to be secured by a stitched whipping, as seen to the right side in the picture.

Method with brummel splice—this splice is self-locking, which means the eye is secured and the end cannot pull out.

A brummel splice has first tucked the short end through the standing end, and then the standing end has been tucked through the short end. These two tucks are made first, and afterward the end is spliced inside the rope as before. There are two ways of making a brummel splice, one by actually tucking the whole length of the standing end through the short end, and another method, yielding the exact same result, but without having to pull the entire length of the standing end through.

BRUMMEL SPLICE BY PULLING THE STANDING END THROUGH

VIDEO 1.2

1

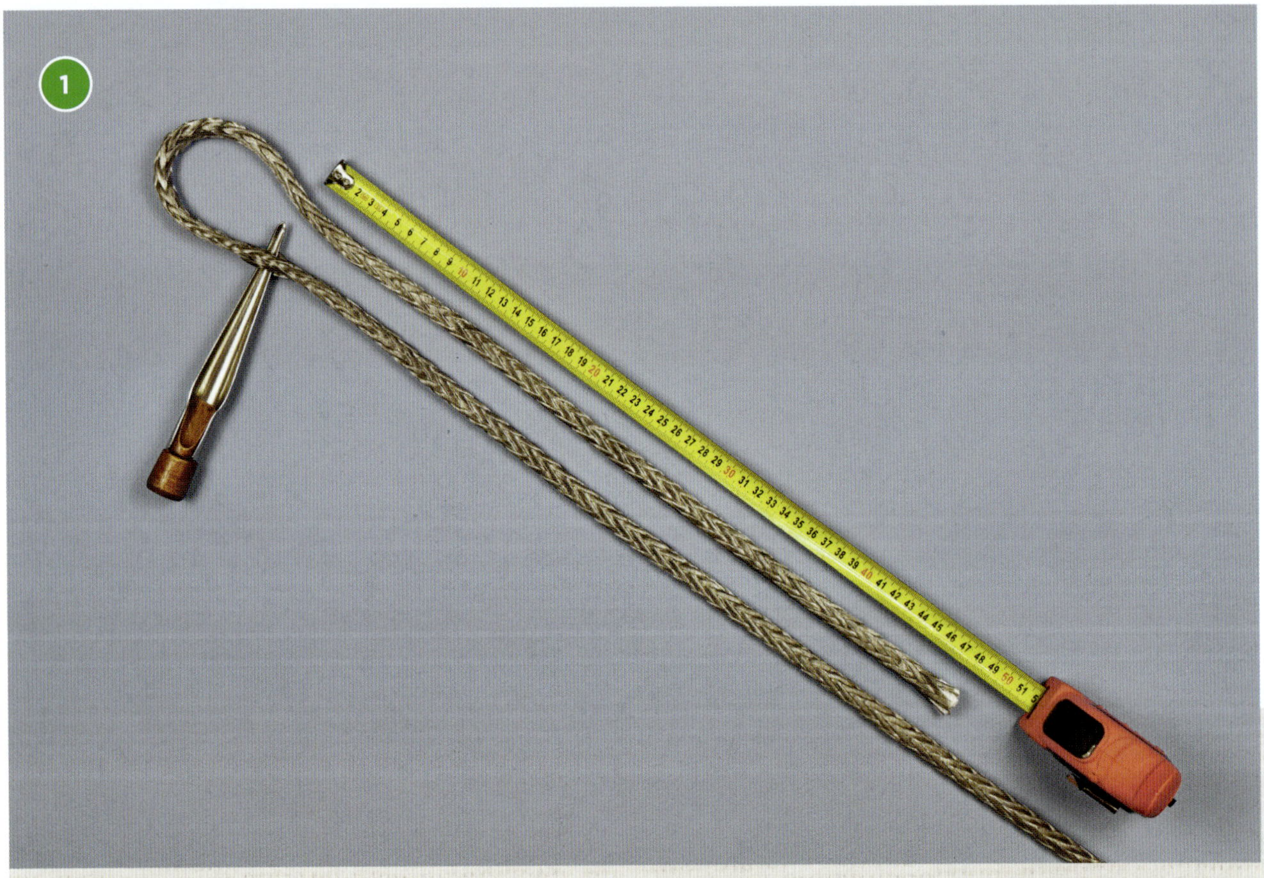

Measure the splicing length. With HMPE ropes, this is 48 times the rope's diameter. Here 48 x 10 mm diameter = 480 mm or 48 cm.

Form the eye in the desired size.

Beginning and end of the eye circumference may be marked with a felt pen, although this is not really necessary after you›ve had more practice.

Using a fid or, ideally, a Swedish fid, poke through the middle of the braid at the beginning of the eye circumference. Poke through between whole strands; do not split any strands when poking through.

Tuck the short end through the widened opening. This is the first of the two tucks of the brummel splice.

Pull the short end through until the eye has the desired size. Then poke close to the first tuck through the short end, using a fid or, ideally, a Swedish fid. Now tuck the standing end through the widened opening in the short end. This is the second tuck that makes up the brummel splice.

Pull the standing end all the way through. Now you have the two tucks that lock the eye. One time it's the short end through the standing end, and one time it's the standing end through the short end. And that's the brummel splice.

Now the short end can be spliced inside the rope as before. Taper the last third of the short end before letting it disappear inside (see method without brummel splice).

The finished splice, including a brummel splice. The strength and extreme holding power come solely from the end being spliced inside the rope over a length of 48 times the diameter. The brummel splice only locks the eye and secures the splice.

BRUMMEL SPLICE WITHOUT PULLING THE STANDING END THROUGH VIDEO 1.2.1

This is a very handy method. Once it's mastered, you will want to work only this way. And quite often, it is not even possible to pull the standing end through, because it may be permanently attached somewhere or may have a thimble or other fitting spliced to its end, or you may be working on a tapered end, and farther away the standing end is much larger in diameter. In cases like that, the standing end is simply too thick to be pulled through, and the following method is what's called for.

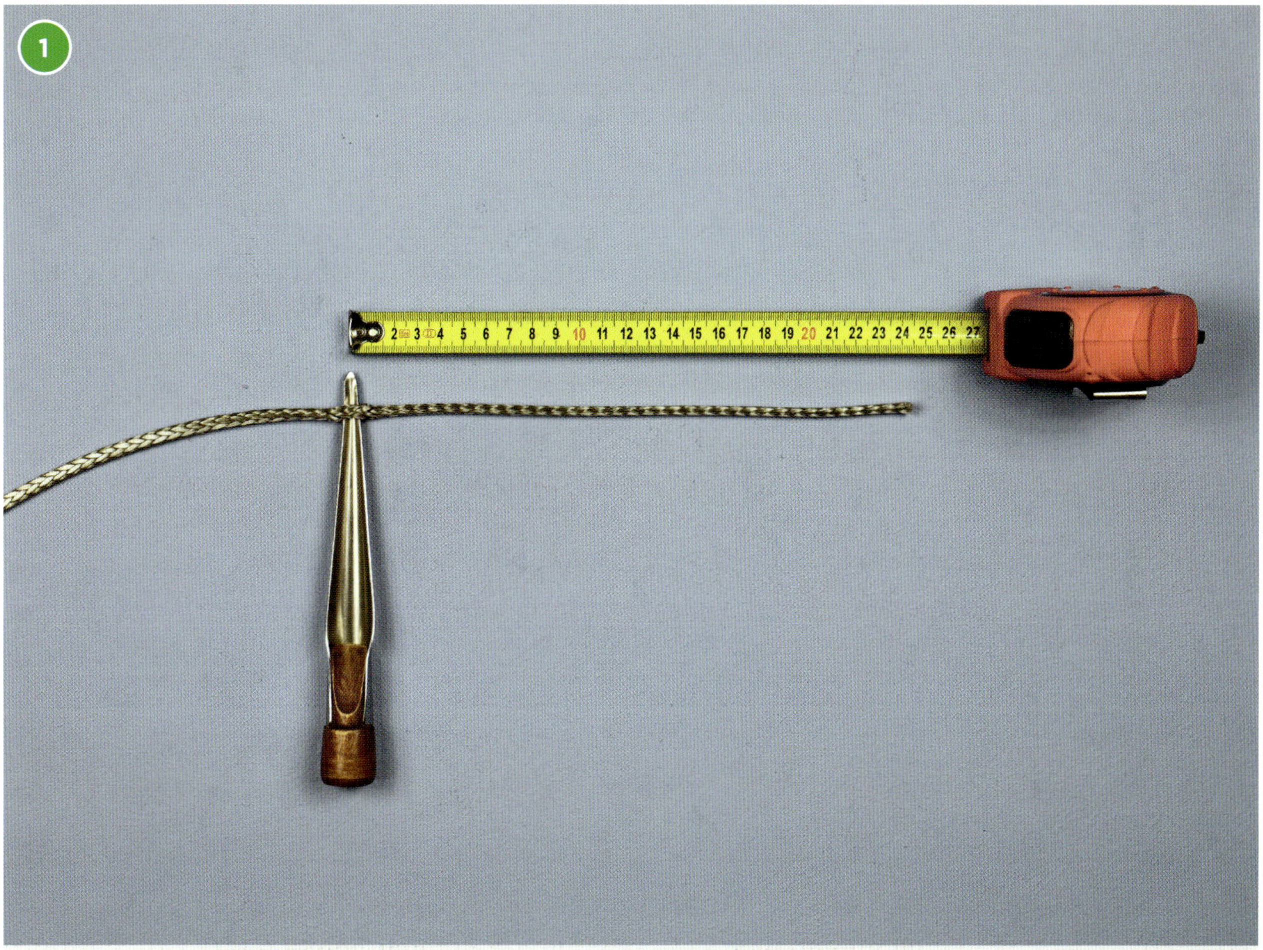

Measure the splicing length. With high-modulus fiber ropes, this is 48 times the rope's diameter. Here, 48 x 5 mm diameter = 240 mm or 24 cm. Using a fid or, ideally, a Swedish fid, poke through the middle of the braid at the measured distance that marks the beginning of the eye circumference. Poke through between whole strands; do not split any strands when poking through.

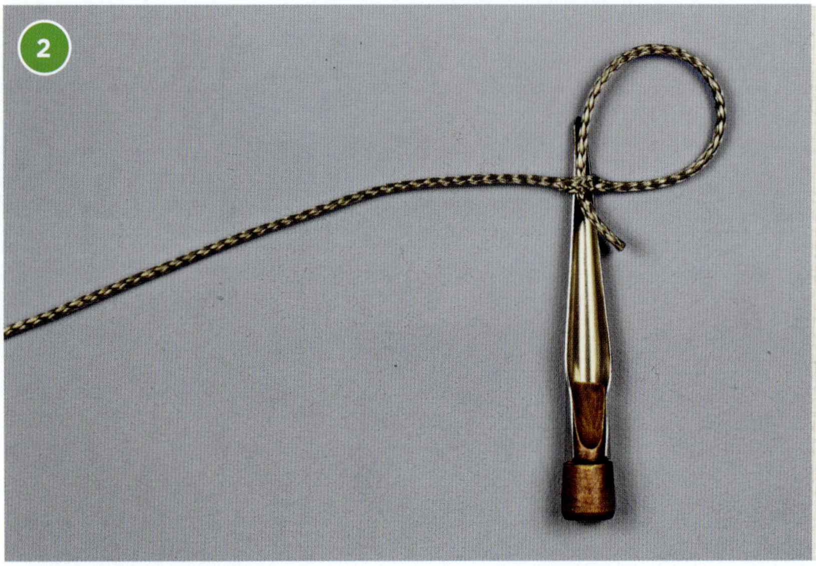

Tuck the short end through the widened opening from the top down, as shown.

Pull the short end through all the way until …

… the braid "capsizes" and only a small opening remains visible. It may be necessary to help the braid along by hand to flip over completely.

The capsized spot marks the beginning of the eye circumference. In case you want to make a thimble splice, lay the rope around the thimble as shown in order to transfere the circumference of the thimble. Be careful not to induce any twist into the rope.

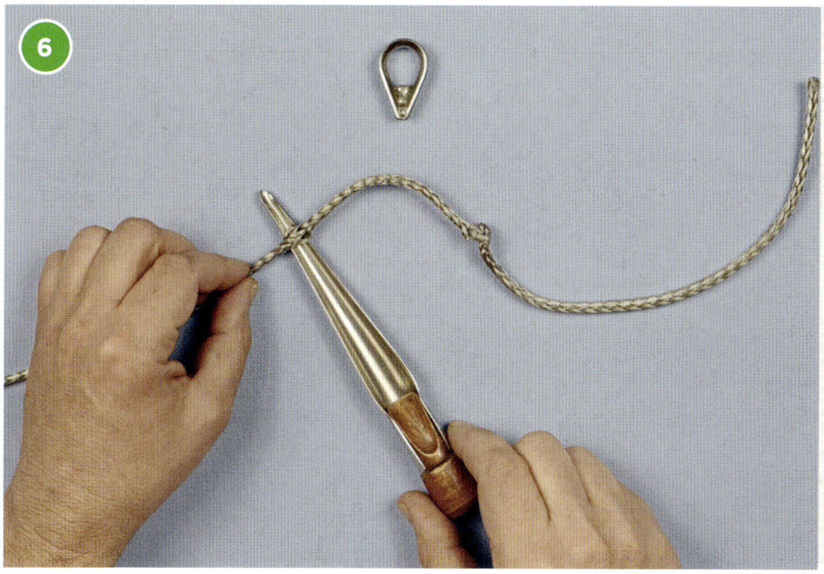

At the end point of the required circumference, poke again exactly through the middle of the braid.

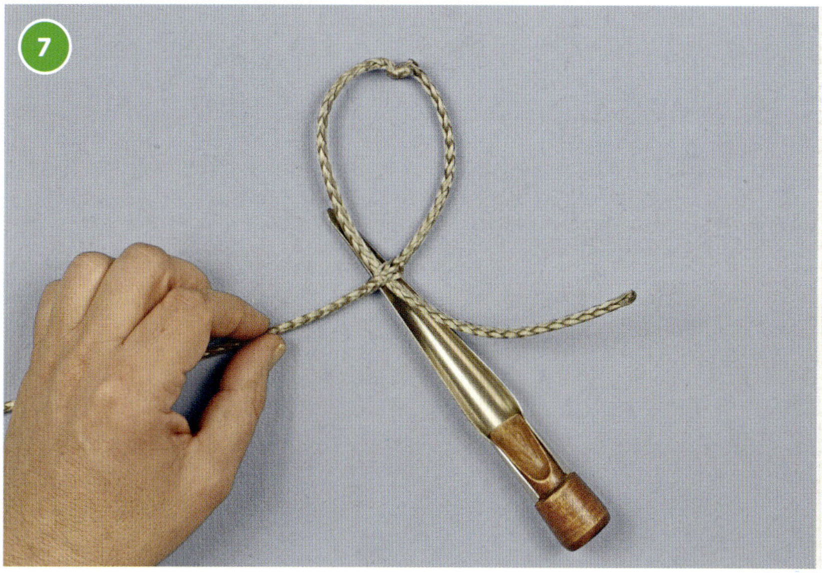

Again tuck the short end from the top down through the widened opening. Be careful not to twist the rope.

Pull the short end, including the first capsized spot, through. Pull it through all the way until …

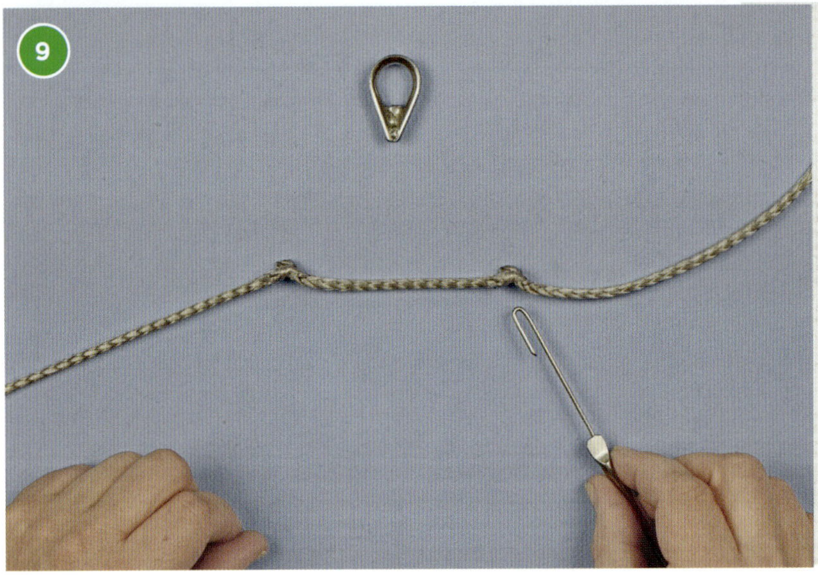

… the braid capsizes. The two capsized spots mark the beginning and end of the eye circumference. Using a small wire hook or similar tool, poke through the first capsized spot from below.

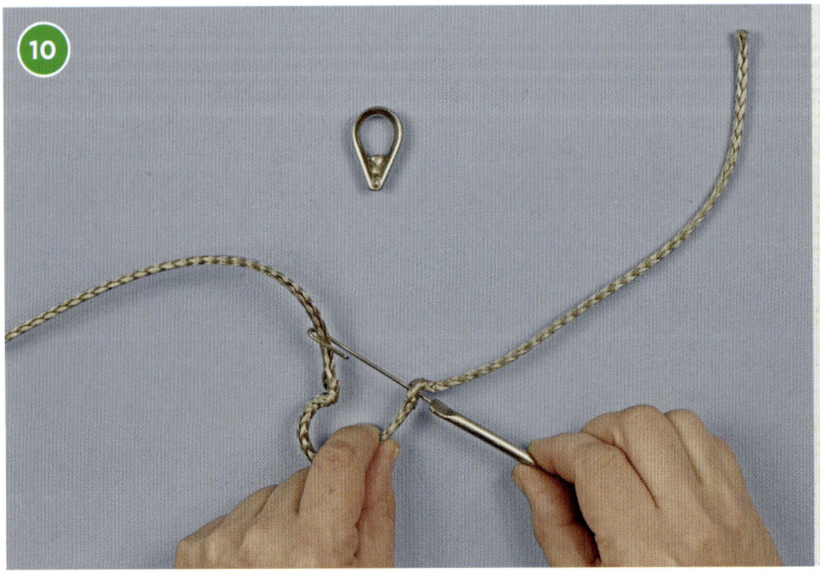

Take the standing end onto the hook and …

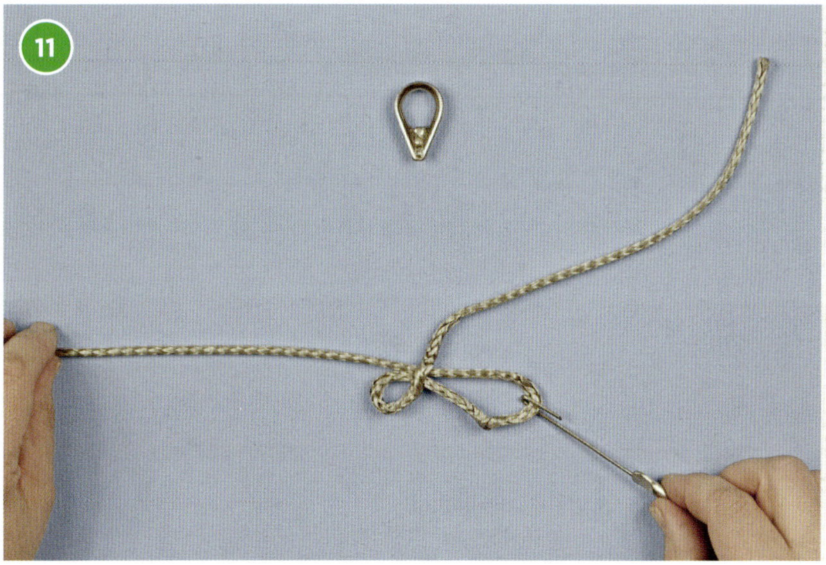

... pull a bight of the standing end through the capsized spot.

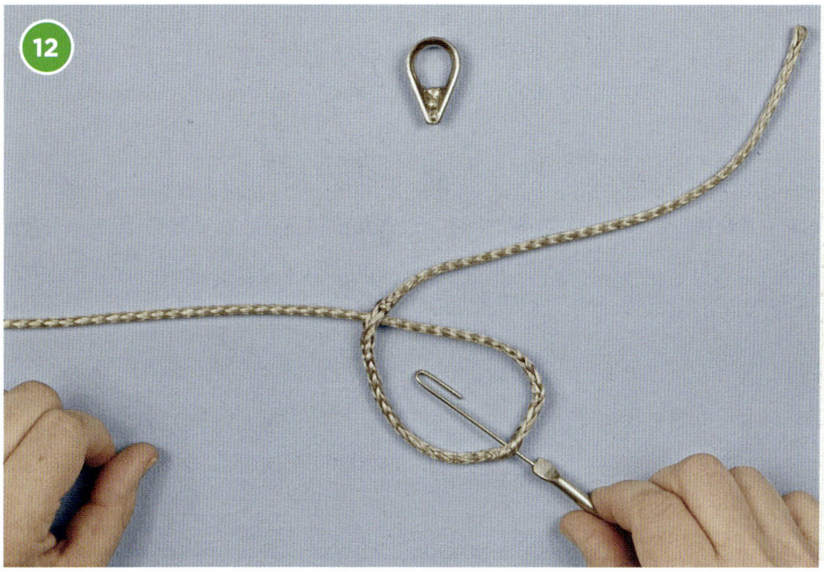

When the bight is pulled through completely, the capsized spot flips back, and the rope's braid is entirely unscathed and just as it was before.

Be especially careful not to twist the rope. Take the small hook and poke through the second capsized spot, going from the outside to the inside of the eye that has formed.

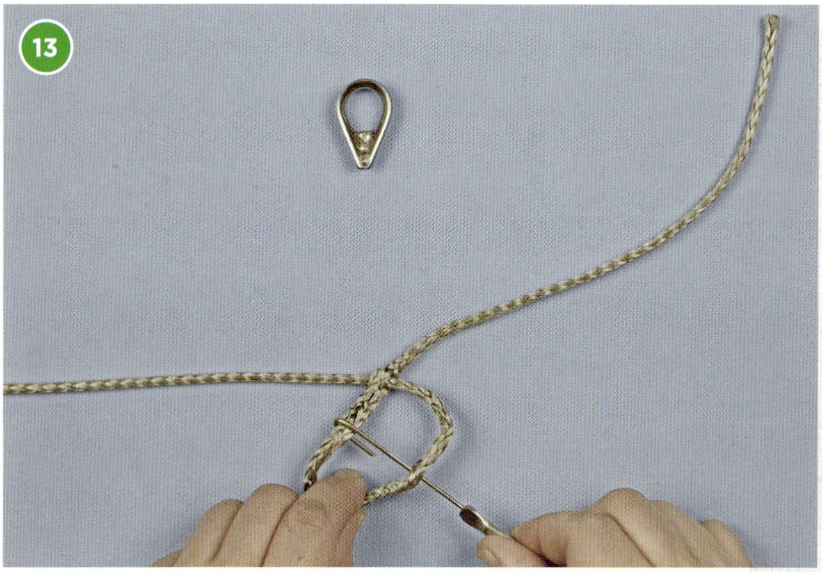

Take this section of the rope onto the hook, which lies between the two capsized spots.

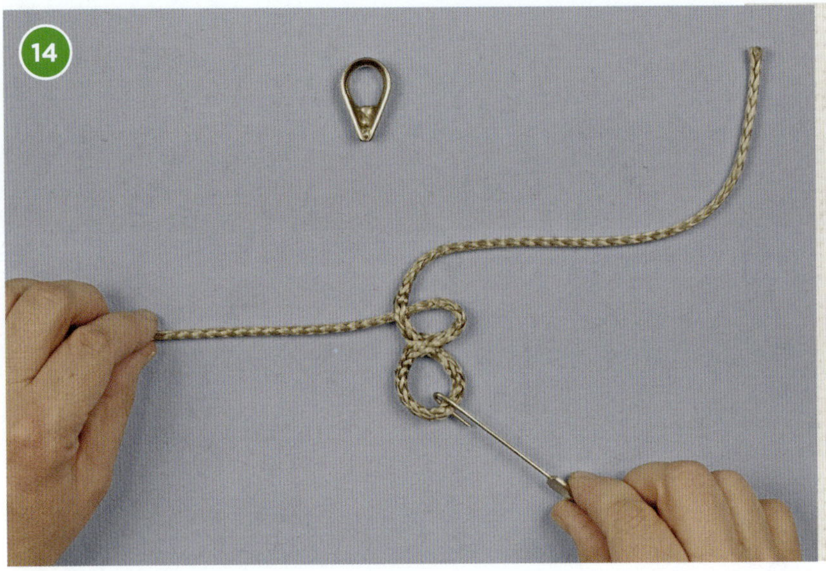

Pull this section through completely, and the second capsized spot also flips back. Again, the rope's braid is entirely unscathed and is just as it was before. Now the two locking tucks of the brummel splice have emerged, but this time it was not necessary to pull the whole length of the standing end through.

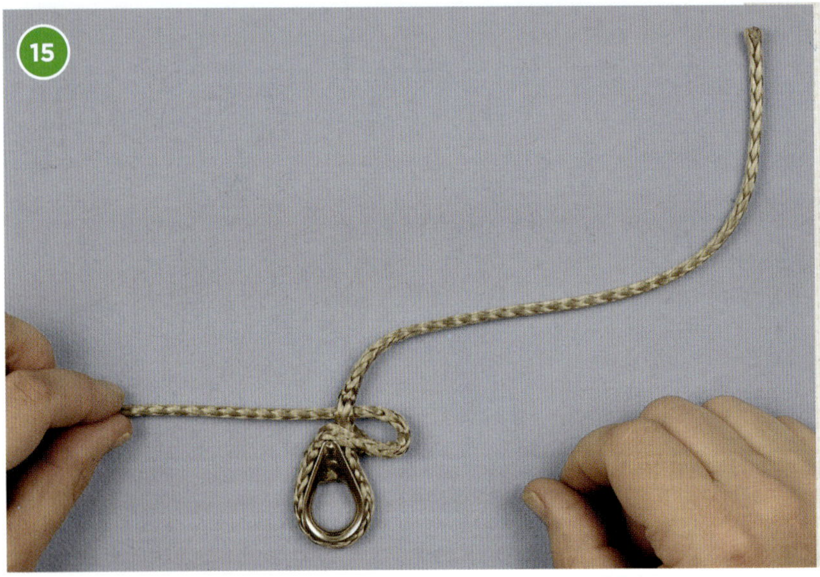

It is easier to insert the thimble into the eye before pulling the first tuck completely tight. The thimble should just fit into the eye to be held very tightly.

Pull on the standing end to close the brummel splice completely.

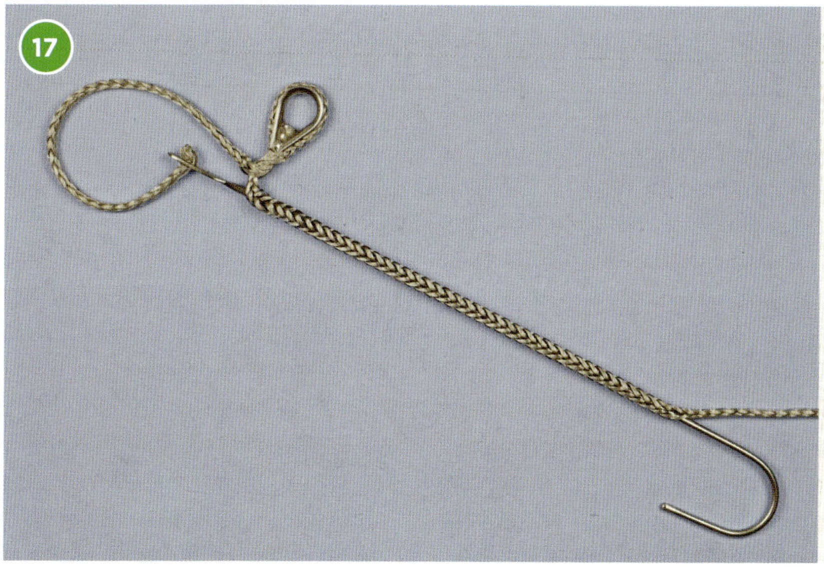

Now the short end is spliced into the rope as before and tapered over the last third of its length (see method without brummel splice).

The finished thimble splice, including a brummel splice. The holding power and the extreme breaking strength come from the short end being spliced into the rope over a length of 48 times the rope's diameter. The brummel splice only serves to secure and lock the splice so that the eye cannot pull open.

VARIATION

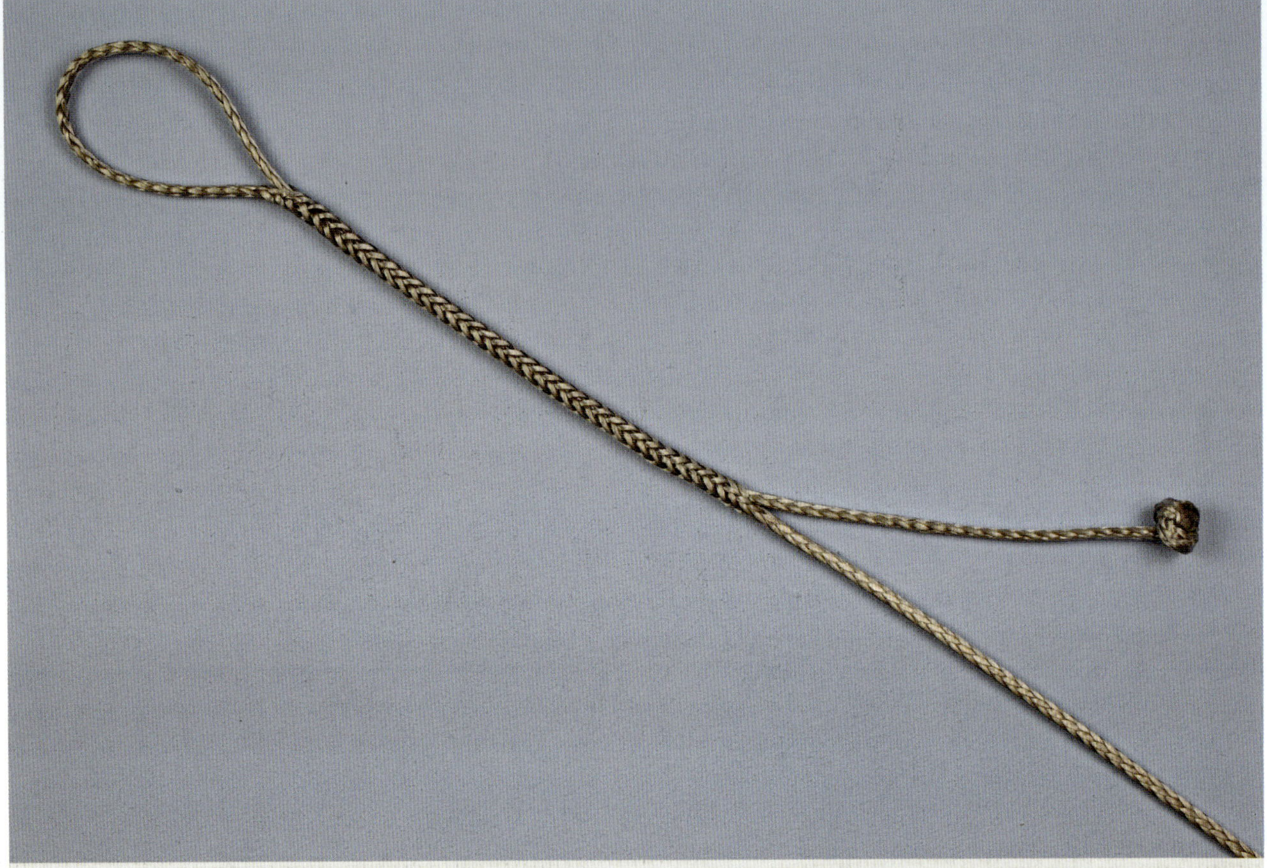

An adjustable eye can be made if a longer end is spliced into the rope over 48 times the diameter and allowed to exit after the splicing length. It›s best to make a stopper knot at the very end. The eye splice can be opened and closed when there is no significant load on the rope, which allows the length of a strop to be adjusted in various applications. When the rope is under load, the splice jams and holds.

Applications: adjustable bridle for the main sheet on small sport boats, hoisting slings, and dinghy trapeze lines from 1.5 to 3 mm Dyneema (instead of wire), which do not need constant adjustment, etc.

NOT SO GOOD

Sometimes you may see splicing instructions where the short end is merely tucked back and forth through the standing end before being spliced into the rope. This does not fully secure the splice. There should be two opposing tucks, as shown before, to create a self-locking brummel splice.

Typical result after strength-testing an eye splice in single-braided high-modulus rope. The rope breaks in the area where the tapered short end terminates inside the rope, and where the diameter changes from thickened in the splicing length to thinner in the rope itself. This 5 mm Dyneema rope broke with a loud bang at a loading of 24.16 kN, which is 2.46 metric tons or 5,423 lbs. This loading corresponds to 93% of the rope's nominal breaking strength of 26 kN. The more careful the taper is made, the higher the breaking strength will be. The soft shackle, made of the exact same 5 mm Dyneema rope, had no problems holding the load.

1.3 JOINING ROPE SPLICE IN CONVENTIONAL AND HIGH-MODULUS FIBER ROPES

Single-braided ropes (without core) made of conventional fibers such as polyester or made of high-modulus fibers such as Dyneema, Spectra, Vectran, and others.

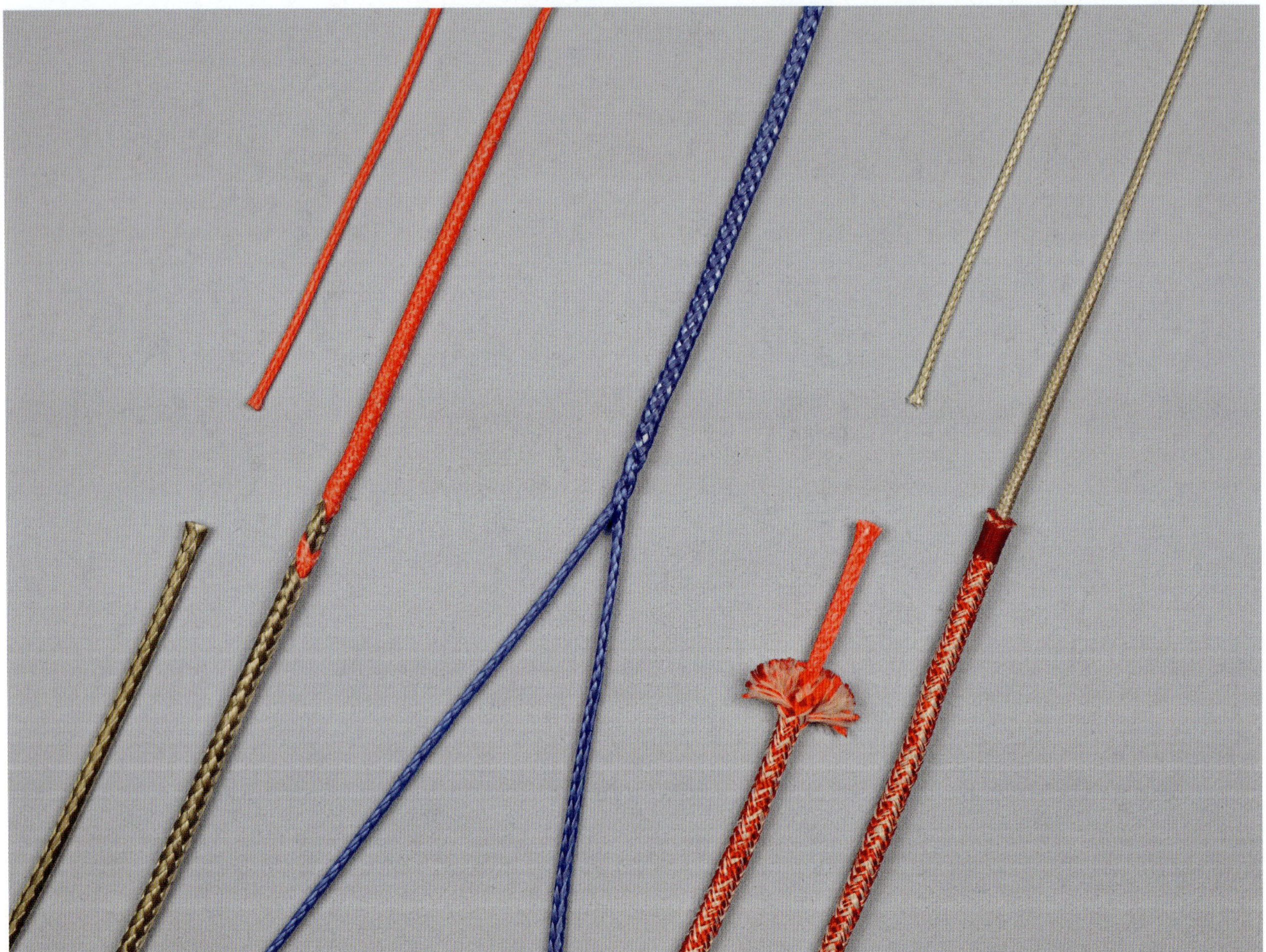

Two ropes can be joined to each other or splicedpoor break together in the form of a "Y" so that two ends fork off from one rope. Contrary to an end-to-end splice, where a closed loop is formed with one and the same rope, in a joining rope splice it is possible to make the splice self-locking using the technique of a brummel splice. This obviates the need for a whipping or stitching to secure the connection. It is also possible to join ropes of different diameter. By splicing a thin Dyneema rope to the thicker Dyneema core of a double-braided rope, a very significant reduction in diamter can be obtained, saving weight and windage in specialized spinnaker or gennaker sheets for light wind conditions.

START

VIDEO 1.3

Construction:	single braided (without core)
Material:	high-modulus fiber (e.g., Dyneema as well as conventional fiber)
Difficulty:	easy
Splicing tools:	splicing needle or fid

1

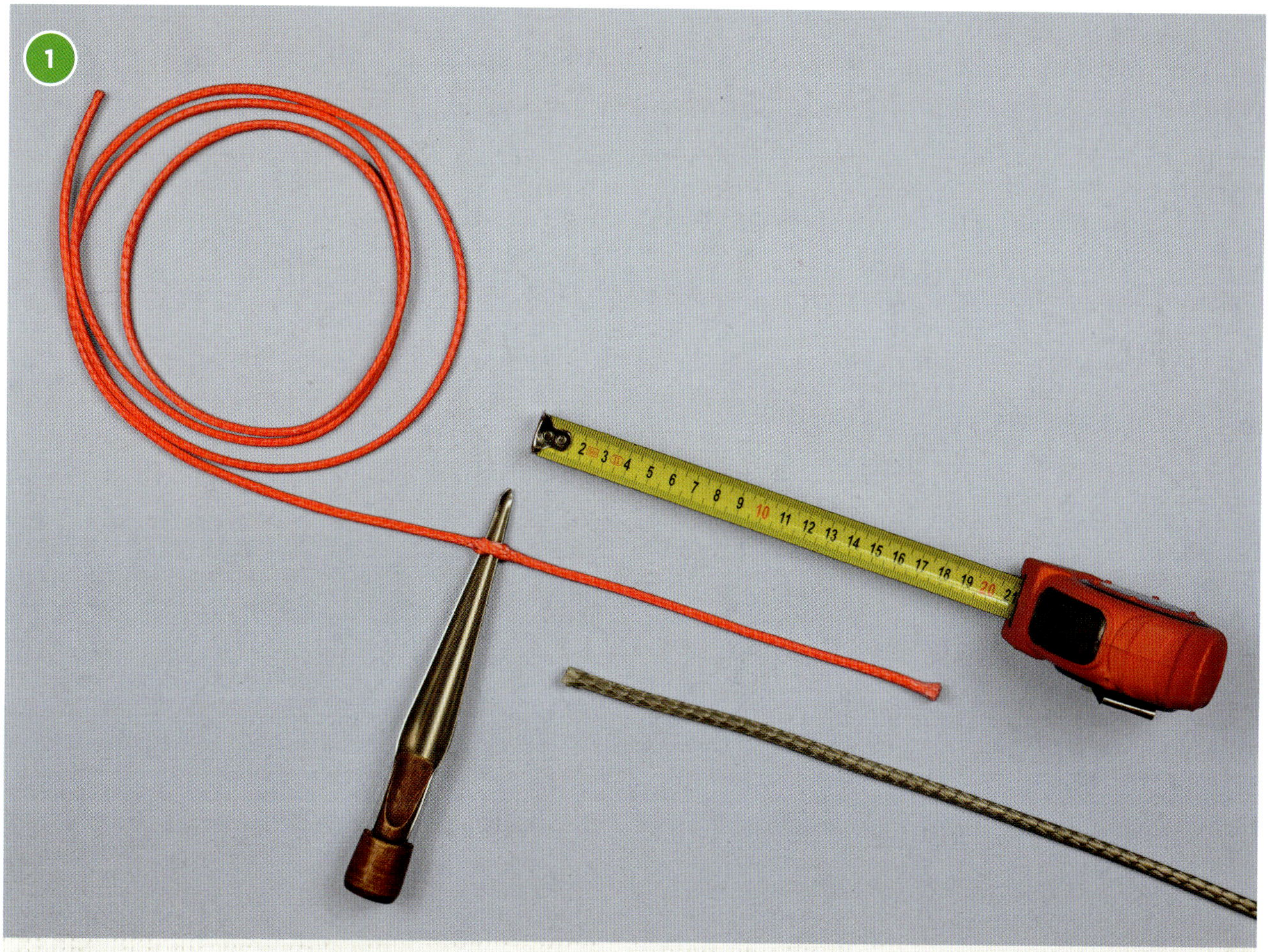

Here a 4 mm and a 6 mm Dyneema rope are to be joined. One of the standing ends of the two ropes will later have to be pulled through the other rope with its whole length. Decide now which of the standing ends of the two ropes you prefer to pull through at a later stage. In this case it will be the thinner, red rope. Measure on this rope a distance of 48 times the rope's diameter from the end. Here, 48 x 4 mm = 192 mm or, rounded off, 19 cm. (When joining ropes of different diameter, all measurements are referred to the thinner diameter.) Poke through the rope at the measured spot, using a marlinspike or, better, a Swedish fid. Widen an opening between the strands. Do not split strands when poking through the braid. Tuck the end of the other rope (the gray rope) through the opening.

With ropes made of polyester or other conventional fibers, the length to be measured is only half as long; namely, 24 times the diameter.

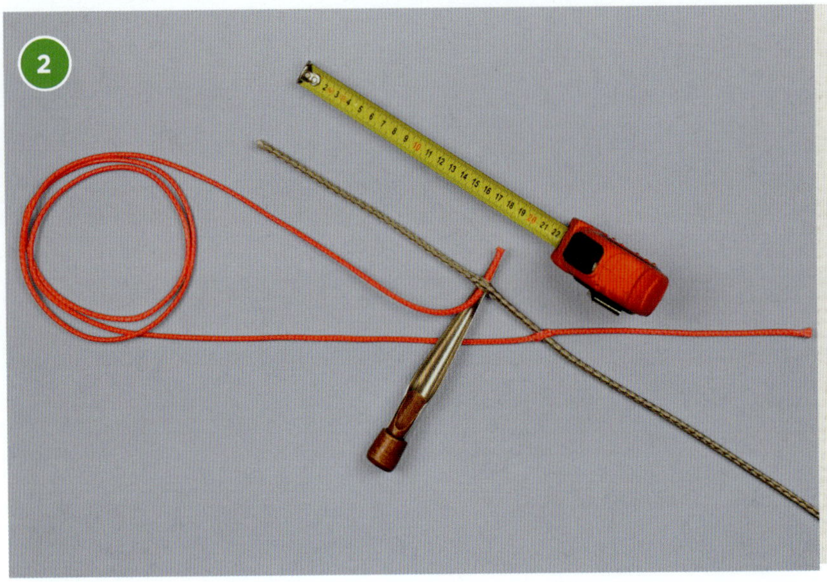

Measure from the end of the rope you just tucked through (the gray rope), again a distance of 19 cm.

(Only in case you want to splice a "Y," pull the rope through the opening up to the spot where you want the rope to fork off into two ends. The Y splice will be finished after completing step 5.)

Poke through the braid at the measured spot and widen a small opening. Now tuck the standing end of the other rope (the red rope) through this opening. This gives a self-locking brummel splice.

Pull tightly on both standing ends of the two ropes to close the opposing tucks of the brummel splice. Next, the two short ends will be spliced into the ropes. Insert the splicing needle a little farther down the rope from where the short end comes to lie, to compensate for the shortening effect and to fully bury the short end.

Pull the short end through all the way and taper it over approximately the last third of its length. Details about tapering can be found in the chapter "Basic Splicing Skills" on page 20. Firmly milk the outer braid back, starting from the brummel splice, and the end will disappear inside the rope.

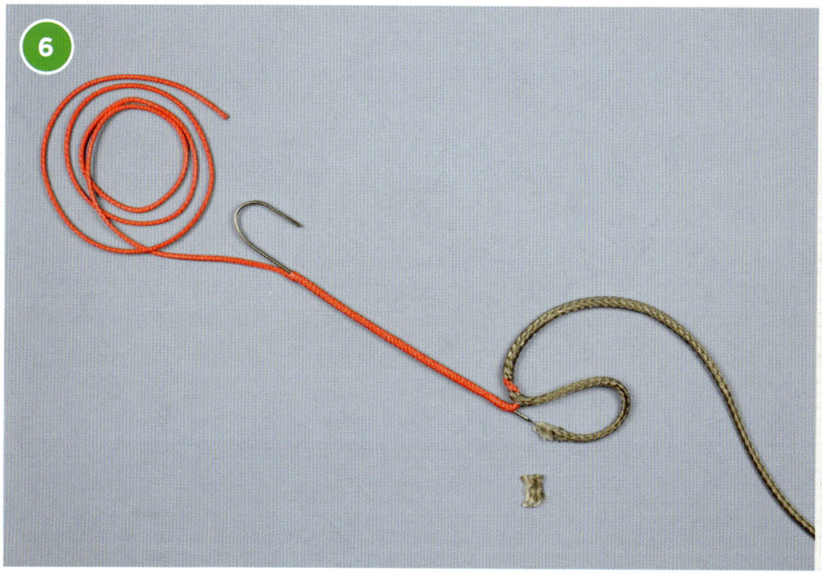

Insert the splicing needle from the other side and exit again very close to the brummel splice. To facilitate splicing the thicker rope into the thinner rope, thread only about half of the yarns onto the splicing needle. Carefully cut the other half of the yarns off. It is one of the great advantages of working with splicing needles that in this way, ends can be attached without any increase in thickness and can be pulled inside much easier.

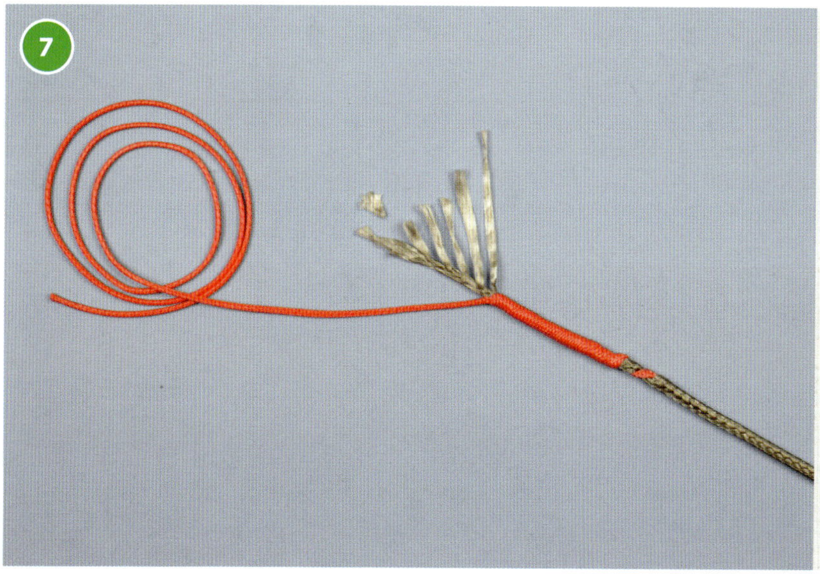

Again pull the short end through all the way and taper the last third of its length. Milk back firmly and the end disappears inside. The joining rope splice is secured by the opposing tucks of the brummel splice and does not need any whipping or stitching. The short ends are compressed under load and generate the holding power of the splice.

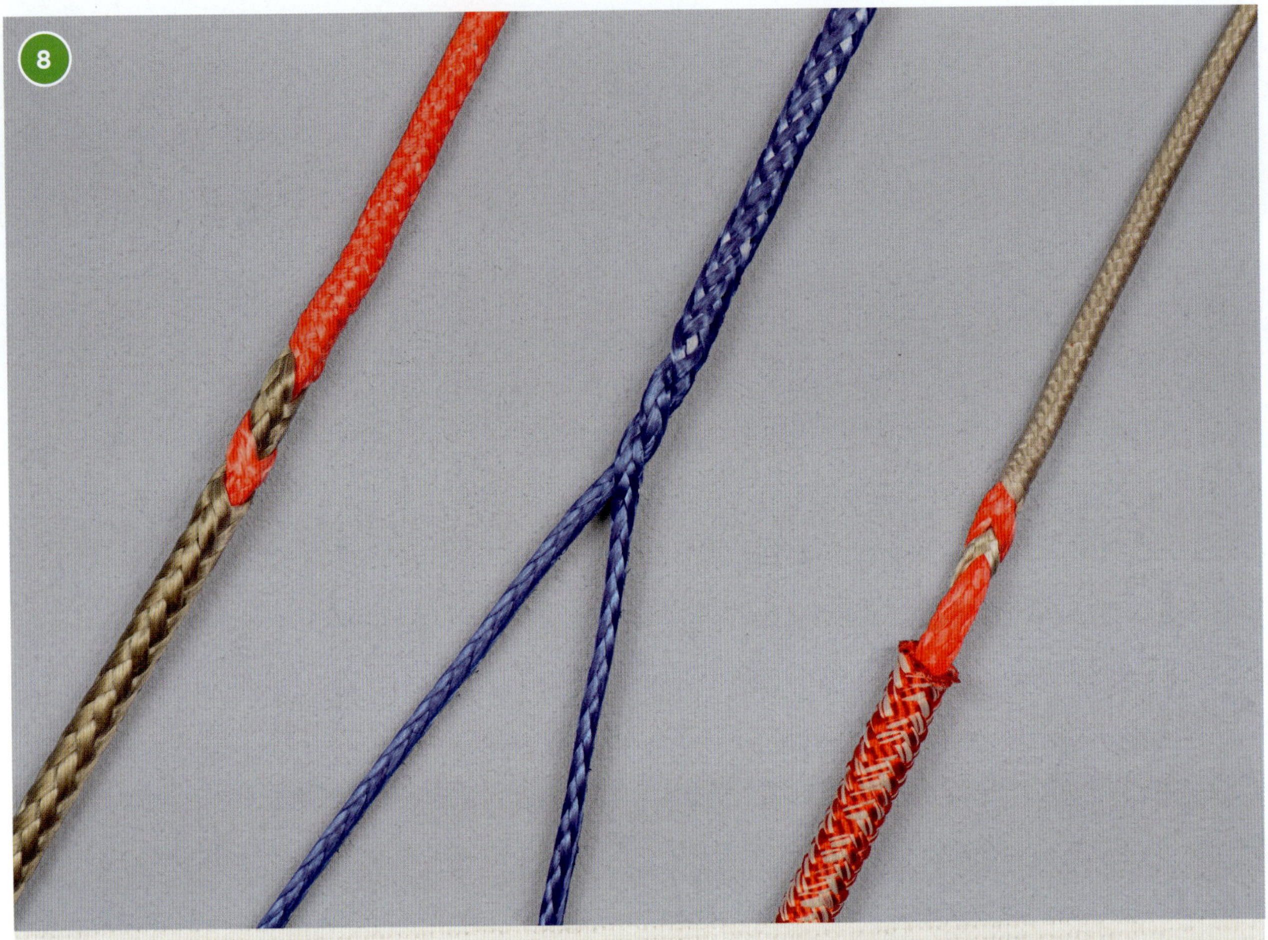

The self-locking brummel splices in joining rope splices. Left: 6 mm Dyneema joined to 4 mm Dyneema. Middle: Y splice in 5 mm Dyneema. Right: 8 mm double-braided rope with a 5 mm core from Dyneema joined to 3 mm single-braided Dyneema. The cover still needs to be milked over the brummel splice and can then be secured with a whipping.

1.4 LOOPS SPLICED WITH DYNEEMA, SPECTRA

Loops from high-modulus fiber rope serve as very strong and lightweight connectors for various fittings and attachments.

Loops are spliced from single-braided high-modulus fiber ropes (e.g., Dyneema, Spectra, and others). Single-braided, mixed-fiber ropes can also be used to create beautiful loops, although the splicing may call for for a little more strength when pulling the ends inside (top left). Thimbles can be seized into loops with strong whipping twine for adjustment of hauling points or sheeting angles. Loops can be doubled or looped through for attachment to fittings.

START

VIDEO 1.4

Construction: single braided (no core)
Material: mixed fiber, Dyneema, Spectra combined with polyester, for example
Difficulty: easy (if the end is tapered carefully)
Splicing tools: splicing needle or similar pull-through tool

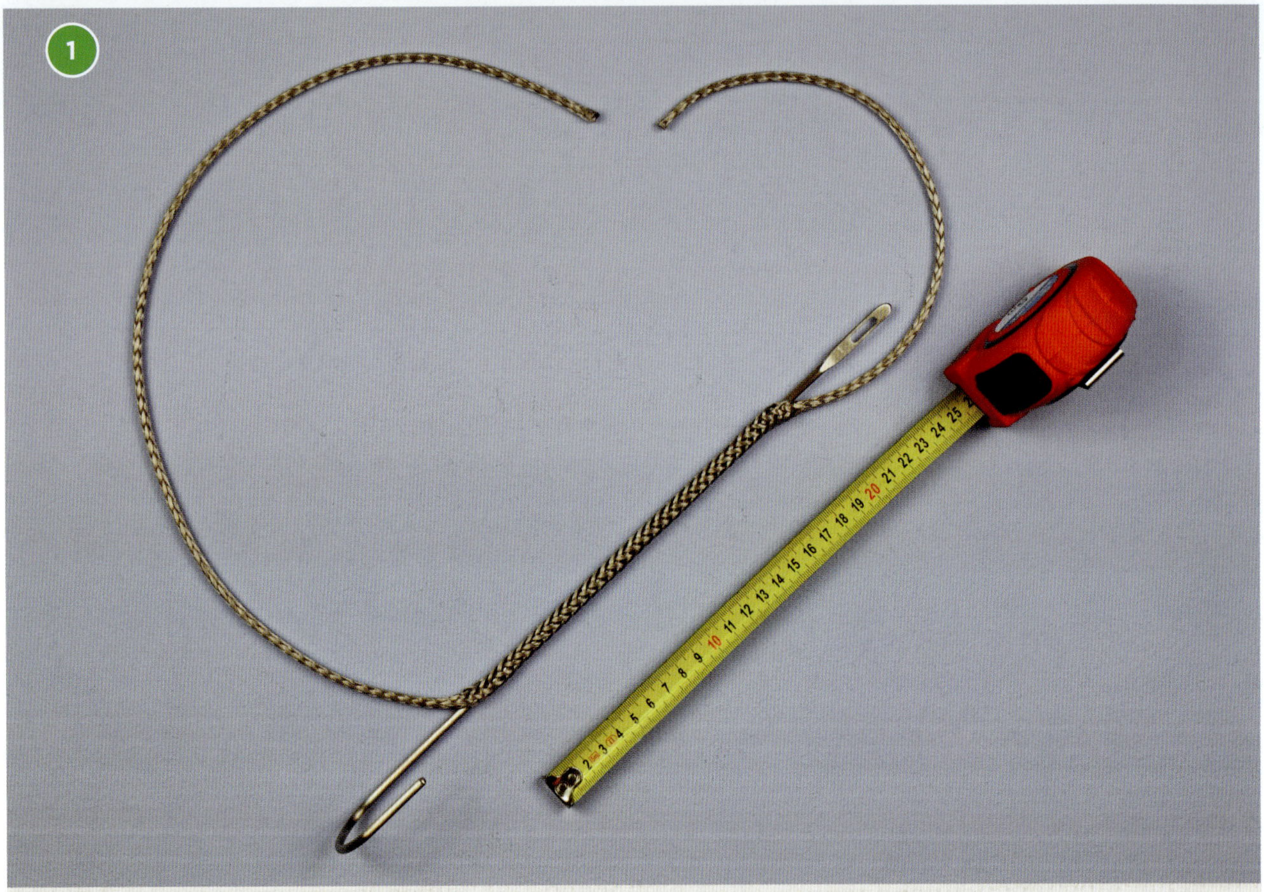

Decide on the desired length of the stretched loop. You will need a little more than five times this length to splice the loop. Here, a loop is spliced from 5 mm Dyneema. When finished, the stretched loop should have a length of 20 cm. That means a rope of about 105 cm length is needed. Insert the splicing needle at the middle of this length at 52.5 cm from the ends. Push the splicing needle toward one side through the braid and over a distance of 20 cm, corresponding to the desired length of the finished loop.

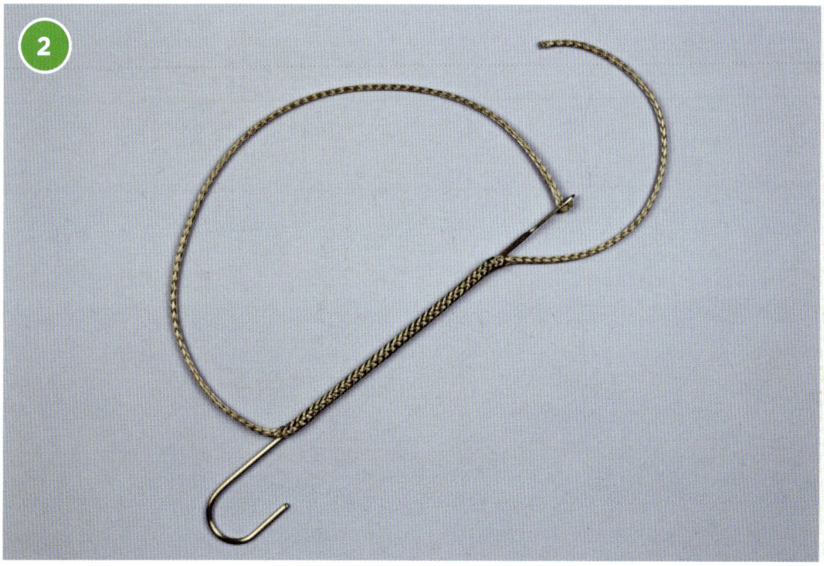

Thread the longer end onto the splicing needle …

… and pull it through until the remaining unspliced section of the loop (here the left side of the loop) is only slightly longer than the spliced section (here the right side of the loop).

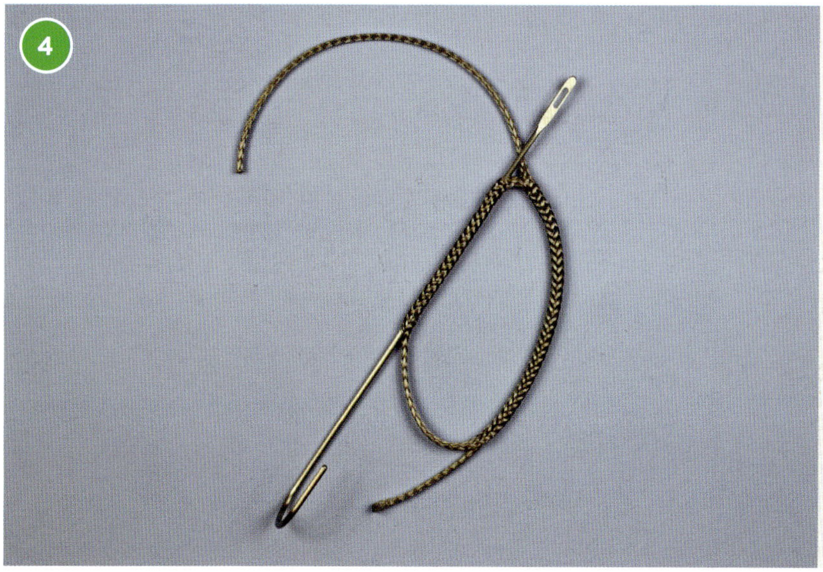

Now the other end is spliced inside, but this is best done in two steps. Insert the splicing needle at a point somewhere in the middle of the unspliced section and exit very close to where the rope disappears inside.

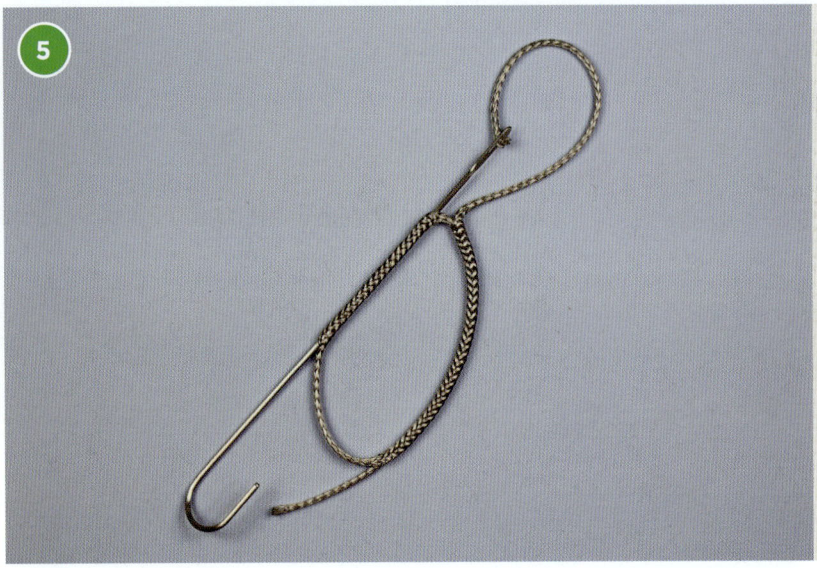

Thread the other end onto the splicing needle and pull it through.

Now insert the splicing needle a few centimeters or about two inches behind the spot where the first end exits. This is the spot to which the second end will be pulled through. The ends will thus overlap each other inside the loop. Be careful not to pierce into the end that has already been pulled inside. The splicing needle must be led between the inner and outer braid.

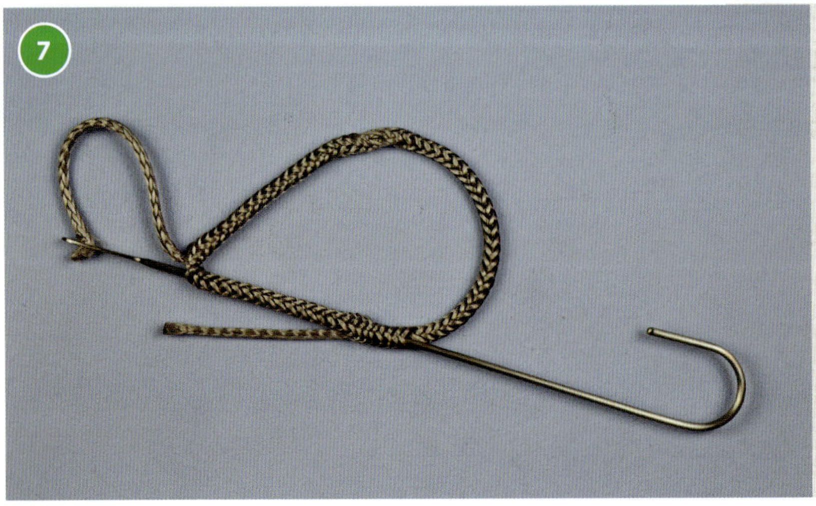

Push the splicing needle through and exit at exactly the same spot where the second end exits. Thread the end onto the splicing needle and pull it all the way through.

Both ends have been spliced inside and overlap each other over a short distance.

Pull on the ends so that the outer braid bunches up and the crossover point is tightly pushed together.

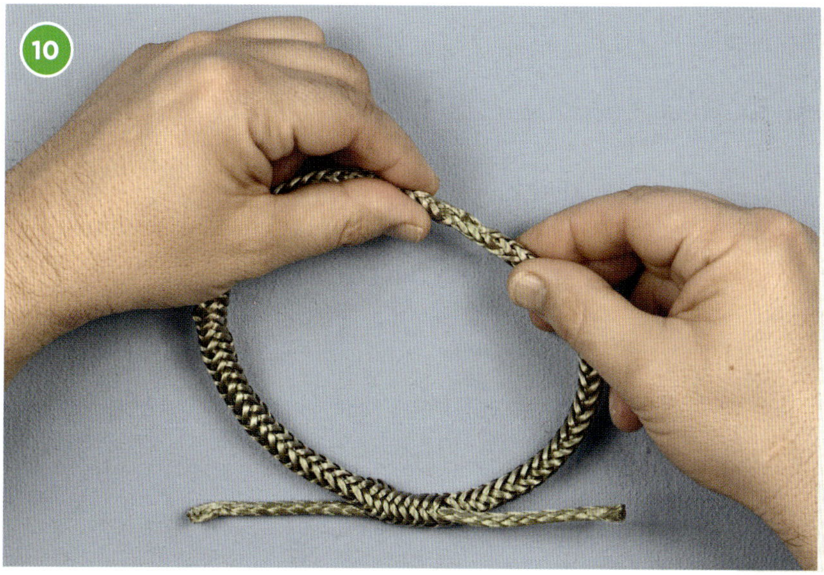

Hold tight at the crossover point and milk the outer braid back firmly to both sides. Stretch the loop repeatedly with the full strength of your hands, so that the yarns of inner and outer braid settle down with no slack remaining.

11

Now the protruding ends are taken care of. Hold one of the ends at exactly the point where it exits, and pull it out just a little farther. Cut it off at the point where you are holding it. Taper the end by cutting it again at a sharp angle.

12

Milk the outer braid back and the end disappears inside. Stretch the loop again with the strength of your hands. Now do the same with the second end.

13

The loop has the same diameter all around, thanks to the overlap of the ends. To finish off, secure the crossover with a stitching. Use whipping twine made of polyester or, ideally, whipping twine in the same color and made of Dyneema.

The same loop made from 5 mm Dyneema after strength testing. Breaking strength was 48.30 kN, which is 4.92 metric tons or 10,847 lbs. Breakage typically occurs where the ends terminate inside and not at the crossover point.

THE DOUBLE-GRAPEVINE LOOP

VIDEO 1.4.1

How to attach an object with a length of rope so that it looks neat and stays put? The double-grapevine loop offers an elegant knotted solution. Breaking strength is not as high as with spliced solutions such as with a loop, but it is still sufficient for many applications. The double-grapevine loop closes tightly around an object and will not slip or slide under load relief.

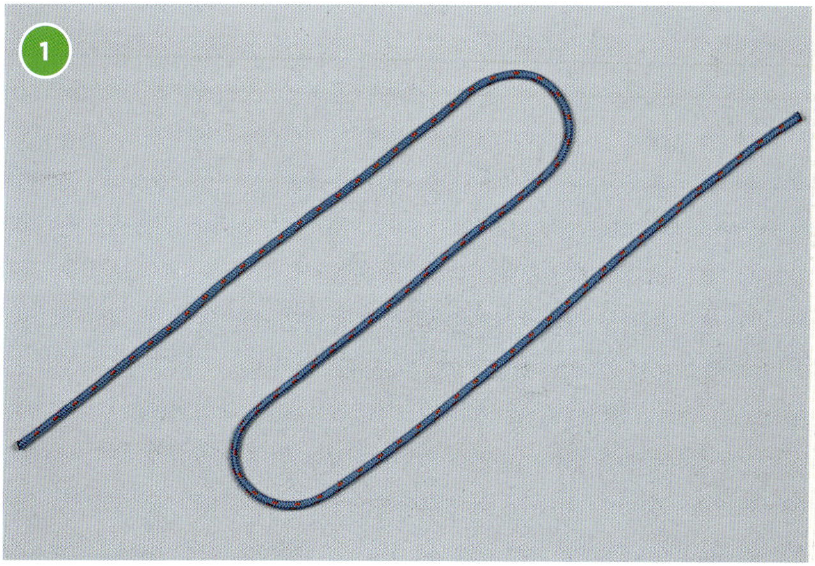

It›s best to use a short length of double-braided rope with a core made of high-modulus fiber (e.g., Dyneema or Spectra). Lay it out in front of you as shown.

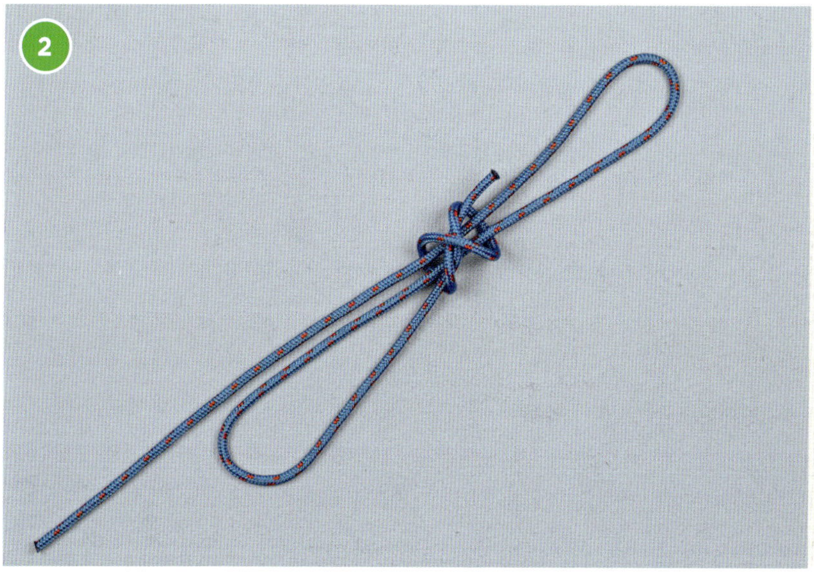

Take one end and tie a double overhand knot around both the other parts. This makes a noose out of the lower bight in the picture and is the same as a scaffold hitch. If required, an object such as a block or fitting can be threaded onto the bight beforehand. Otherwise the double-grapevine loop can be itself looped through an object when finished.

With the other end, tie an opposing double overhand knot around the two other parts, as shown. This makes a noose out of the upper bight in the picture. Again, an object may be threaded onto this bight beforehand if required.

Firmly pull the two overhand knots (or scaffold hitches) tight. The turns of the knots come to lie neatly and snugly next to each other (on the back, the turns cross over once in each knot). Stretch the double -grapevine loop by pulling it with the full strength of your hands. The protruding ends can be cut off very short, which is one of the advantages of the scaffold hitch.

One part of the rope slides through the knots so that the double-grapevine loop can be pulled closed to either side.

It is easy to get things wrong when making this loop, with the result that the loop pulls open under load. Test and make sure before using it in earnest!

Loop one side to an object or attachment point. Pull the noose tight until the turns of the overhand knots (or scaffold hitches) are snug against the object. The double-grapevine loop will stay put. Now the open bight can be used to loop another object through, such as a block.

1.5 SOFT SHACKLES MADE WITH DYNEEMA, SPECTRA

Soft shackles are very strong and lightweight, modern textile connectors that can be opened and closed and can substitute conventional metal shackles in many applications.

Connectors spliced or knotted from old-fashioned twisted rope have been in use already throughout the age of sail. But it is only nowadays with the advent of high-modulus fibers that textile shackles can be fabricated that are so incredibly strong that they not only can be substituted for metal shackles but offer superior, lightweight solutions. Take a look around any modern regatta yacht. You will notice that all attachments are by textile solutions, such as soft shackles, loops, and lashings. Nothing is attached by heavy metal bolts or shackles anymore. A typical application for soft shackles is connecting sheets to the clew of a foresail, such as genoas, jibs, spinnakers, or gennakers. Soft shackles have many advantages: weight-saving by about 90% compared to a metal shackle with equal breaking strength, and no scratching or unnerving cluttering against mast or boom or on deck under load relief. But the best thing about soft shackles is that you can make them yourself in any desired strength and size. All it takes is a few minutes time and a short end of single-braided, high-modulus fiber rope.

START

VIDEO 1.5

Construction:	single braided
Material:	high-modulus fiber (e.g., Dyneema, Spectra)
Difficulty:	easy, once you mastered tying the diamond knot
Splicing tools:	splicing needle or fid

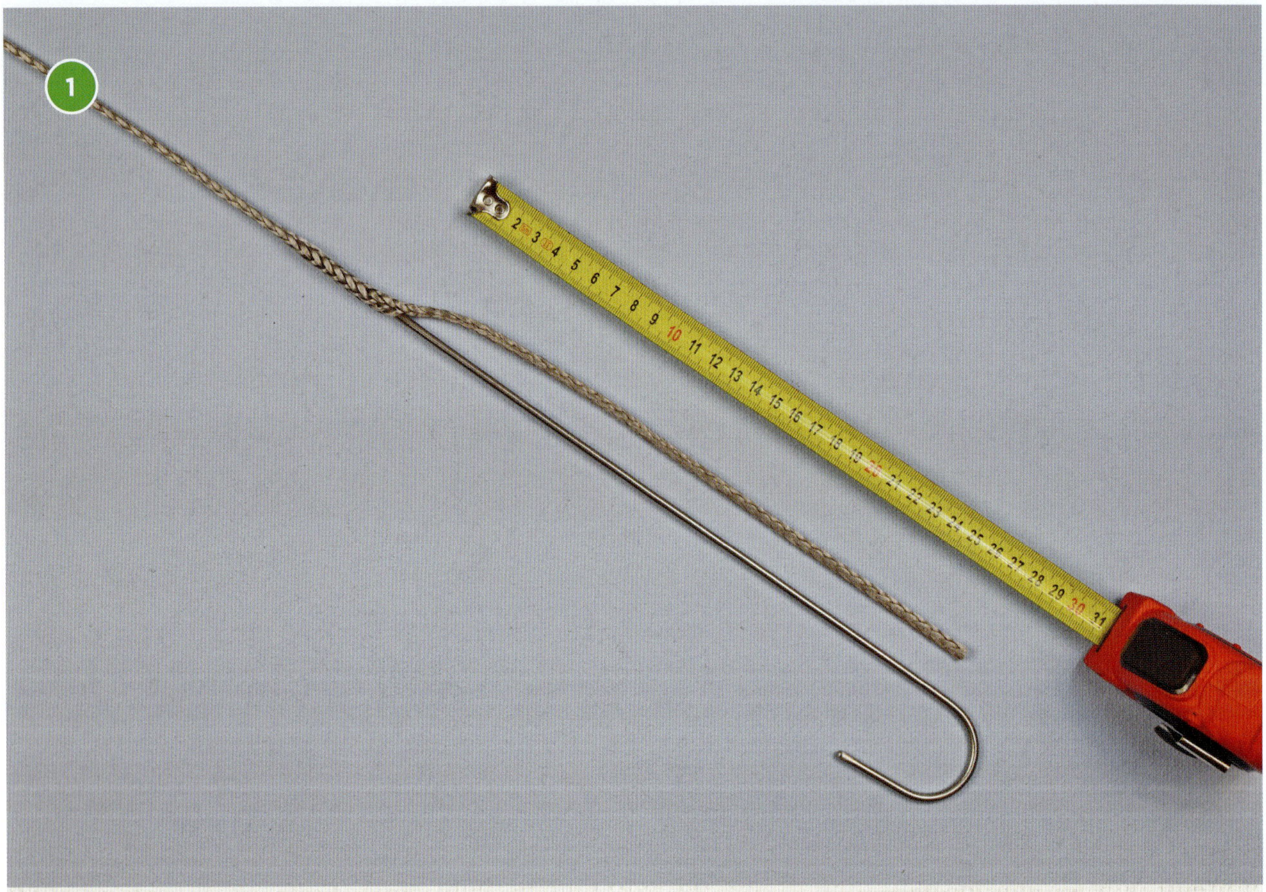

Measure a distance of about 30 cm or 1 ft. from the end of the rope. Insert the splicing needle at this point as shown. Here a 5 mm Dyneema rope is used.

TIP: *The measured distance of 30 cm / 1 ft. is needed to tie and tension the diamond knot at a later stage and is sufficient for ropes up to 6 mm diameter. With ropes of larger diameter, this measurement needs to be increased. As a rule, the measured distance should correspond to at least 50 times the rope's diameter.*

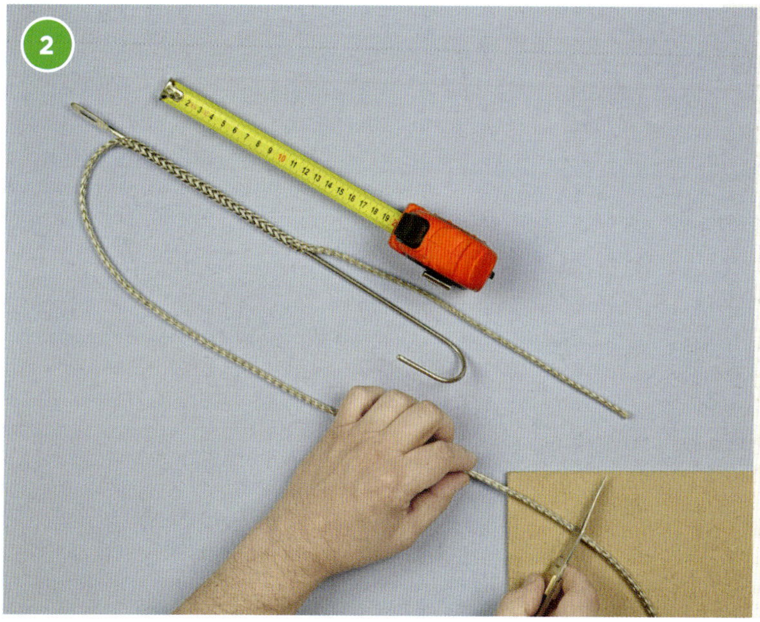

Push the splicing needle along the inside of the rope. Exit with the splicing needle after a distance equal to the desired circumference of the finished soft shackle. Here, the circumference of the soft shackle will be 15 cm. Measure this length and lead the splicing needle out of the inside of the braid, exiting exactly between the strands.

Lead the standing end back parallel to the short end and cut it off at the same length.

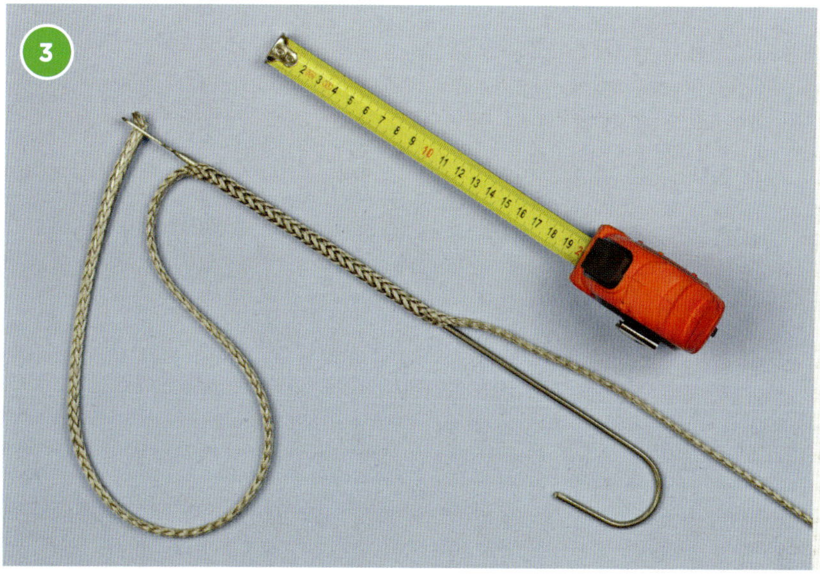

Thread the end you just cut onto the splicing needle and ...

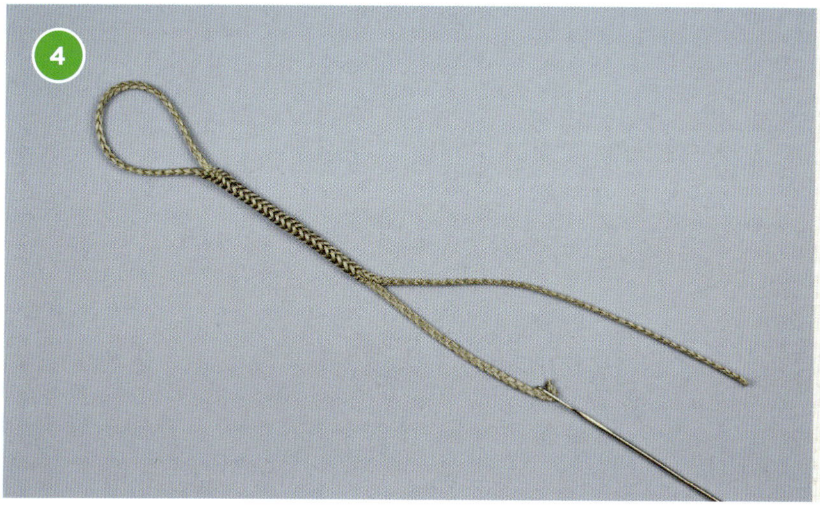

... pull it through to make an eye splice.

Pull the end through until the eye is very small. To be exact, the diameter inside the eye should be equal to the diameter of the spliced section (the section just below the eye where the rope has been pulled back inside itself).

Both ends should now be of about equal length. With these ends, the diamond knot will be made. To be exact again, it's going to be a two-strand diamond knot.

This is the starting position for the diamond knot: Hold the rope in your left hand, between your middle finger and thumb (if right-handed). The eye splice is facing downward. The two ends are also hanging downward: one to the front, the other to the back. (Which of the ends—the one that has been spliced inside or the one that forms the outer braid in the spliced section—is to the front or to the back makes no difference. The diamond knot is completely symmetrical.)

Place your index finger over the end of the spliced section—to the front or to the back makes no difference. The diamond knot is completely symmetrical.)

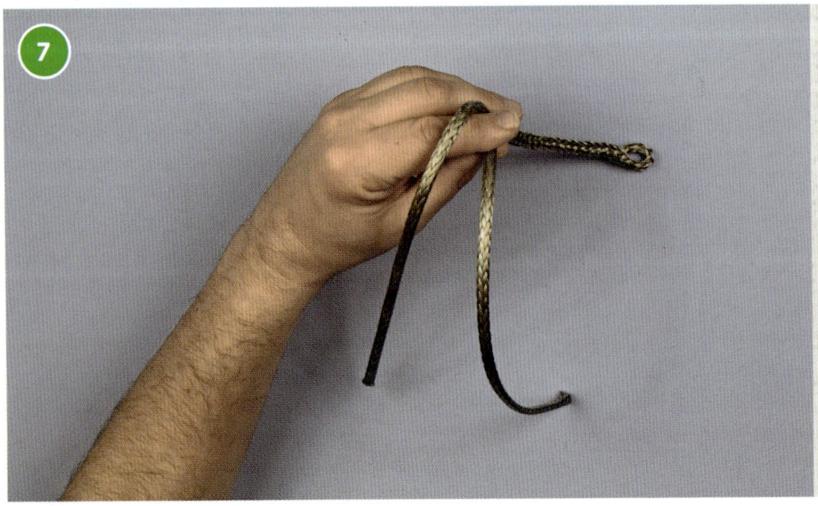

Fold the end that is to the back toward you and to the left of the other end.

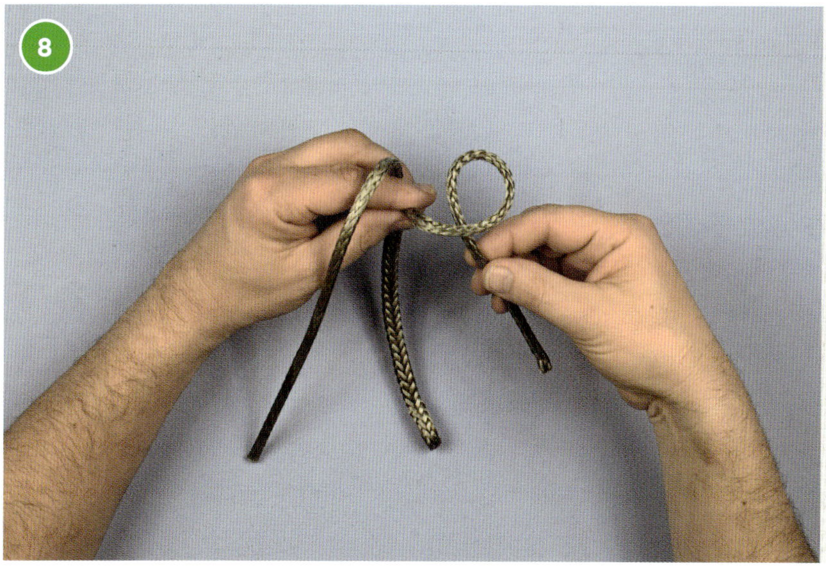

Now take the other end and twist it so that an eye is formed with the short end below the standing end, as shown.

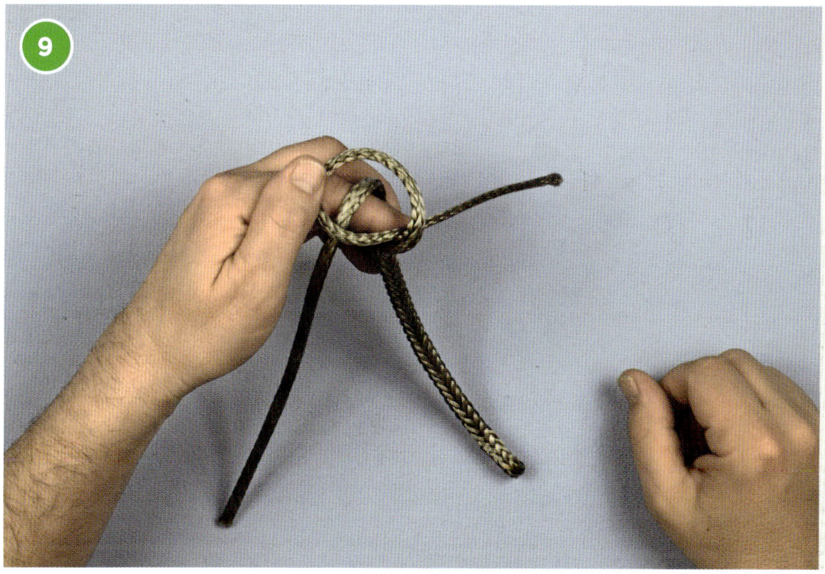

Slide this eye to the left and over the other end, which you had folded toward you in step 7. Hold the eye in this position with your thumb.

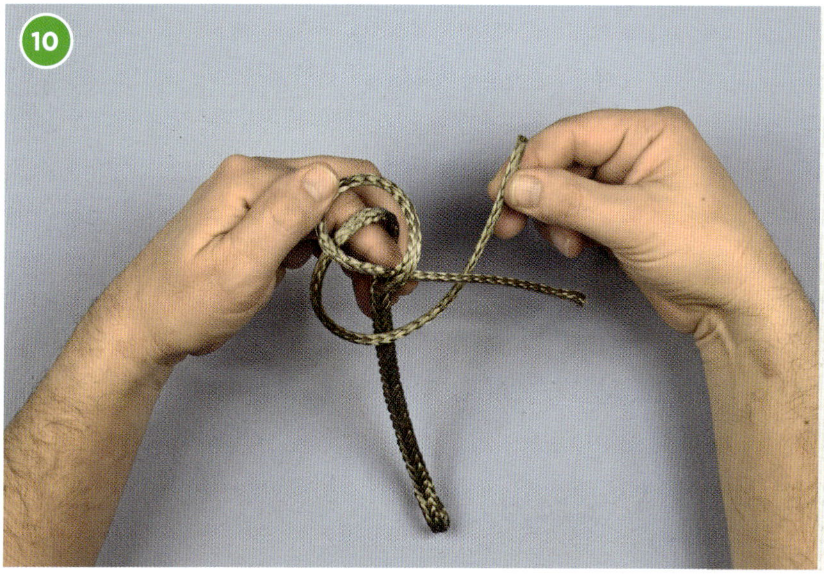

Take the end that you had folded toward you and lead it in front of the spliced section but behind the other end.

Tuck the end under itself and inside the area where you had slid the eye over it. Note that the end is tucked under its own standing end only and stays above the other end that forms the eye.

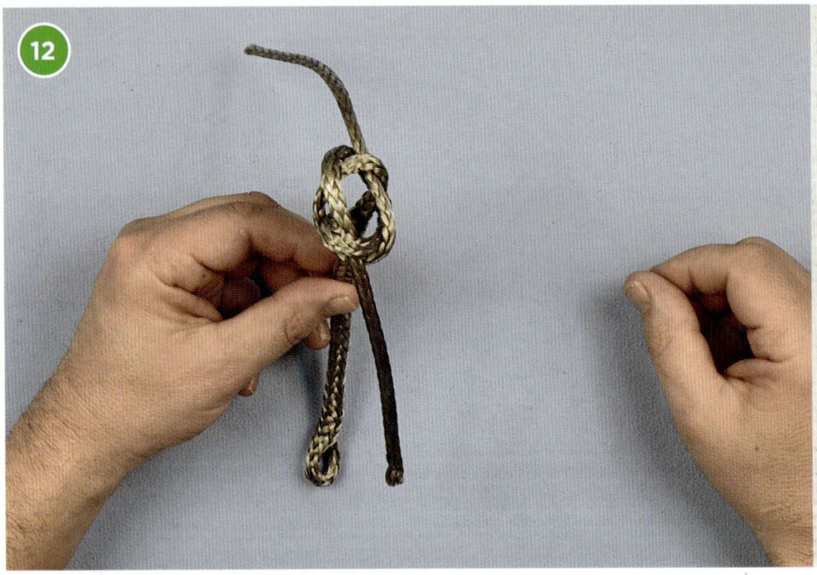

This is what your knot should look like by now, and you can withdraw your index finger. (Incidentally, what you have done so far is nothing other but a carrick bend.)

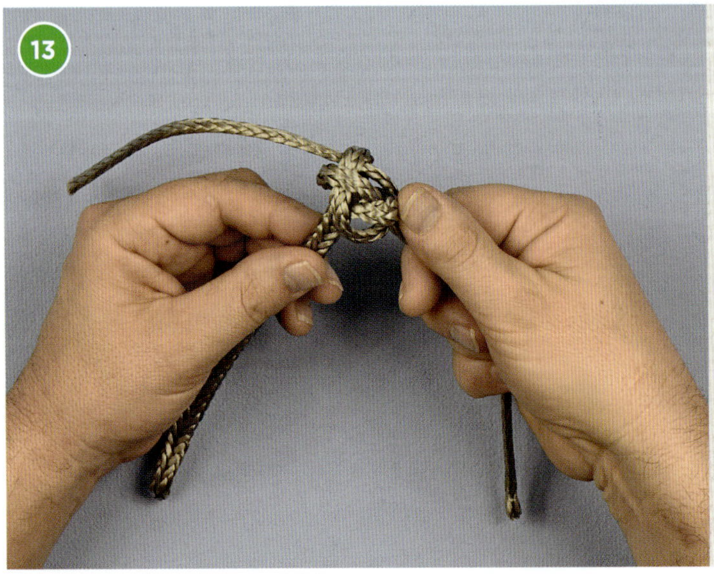

Even out the knot into a regular shape and close it just loosely. To do so, do not just pull on the ends but, rather, use your right hand to push down from above with the thumb and simultaneously take up the slack by tightening from the end. Then rotate the knot by 180° and push down and tighten the other end in exactly the same way. By doing so, the knot can be positioned right at the end of the spliced section.

To finish off, both ends will have to be tucked upward through the middle of the knot. Each of the two ends will be tucked in identical ways. It makes no difference which of the ends is tucked first.

Take one end and lead it around at the bottom of the knot, maintaining the direction of rotation by which the end emerges from the knot. (When looking at it from above, the end pictured here is led around at the bottom of the knot in a counterclockwise direction.)

When leading the end around, go just past the spliced section and immediately after tuck the end upward …

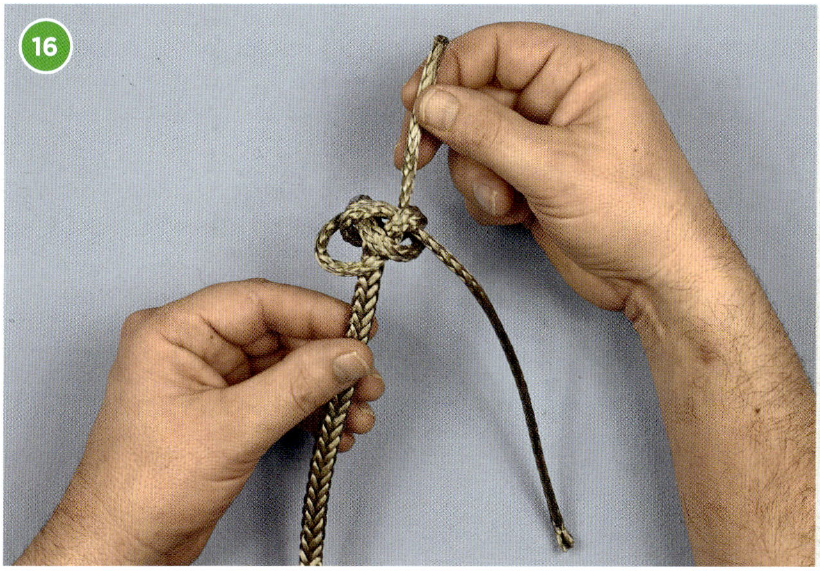

… through the center of the knot.

17

Rotate the knot by 180° and do exactly the same with the other end. Lead it around at the bottom of the knot in the same direction of rotation as the first end, so it is again anticlockwise if looking at it from above. Right after the end is led passed the spliced section it is tucked straight upwards…

18

… through the middle of the knot. Both ends emerge next to each other at the top of the knot. It is easy to tell whether the diamond knot has been tied correctly or not. The diamond knot should show a perfectly regular weave pattern all around.

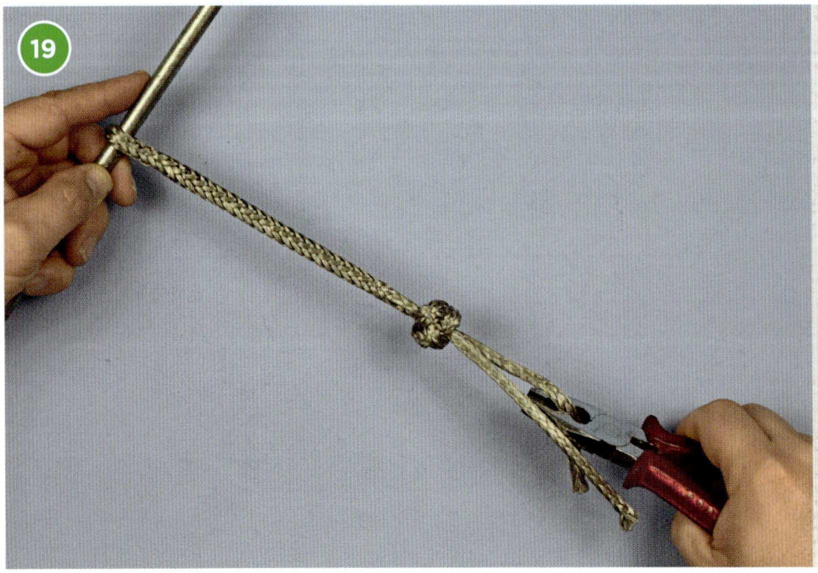

19

In a soft shackle it is important to tighten the diamond knot with your full strength. Tie or hook the eye splice to a strong point or put a fid or screwdriver through the eye splice to be able to really pull strongly at the ends. Use a marlinspike or a pair of pliers to pull the ends separately and with full strength.

The diamond knot has been tightened fully. Note how much smaller it is after being tightened. Now the ends can be cut off short and …

… are fused by a small flame.

Ideally use an electric- or gas-powered heat cutter or soldering iron to melt and fuse the ends just slightly.

23 This is how the fused ends should look, giving the diamond knot a nice rounded shape.

24 Finished soft shackles made from 5 mm Dyneema. The tested breaking strength amounts to an astonishing 29.8 kN or 3.04 metric tons, equaling 6,702 lbs.

25 To open the soft shackle, pull the eye splice open and take out the diamond knot.

To close the soft shackle, put the diamond knot through the eye and lightly milk the outer braid back, starting at the diamond knot so that the eye closes tightly.

Typical breaking strength of soft shackles made from 100% Dyneema SK78:

Made from (rope diameter) 4 mm = 1,440 metric tons or 3175 lbs.
5 mm = 3,045 metric tons or 6713 lbs.
6 mm = 6,020 metric tons or 13.272 lbs.

ATTACHING A DRAW LINE

VIDEO 1.5.1

A draw line allows the soft shackle to be opened more easily and quickly. It is no longer necessary to determine on which side you need to grab the eye to pull it open with your fingers. Suffice to say, simply grab the soft shackle below the eye splice and pull on the little draw line.

Close the eye splice all the way by lightly milking back the outer braid, starting at the diamond knot. Then take the rope between the fingernails of your right hand at exactly the spot where it enters inside.

Pull the rope out over a short distance. Hold it with your left hand, about 2 cm away from where you are holding it with your right hand.

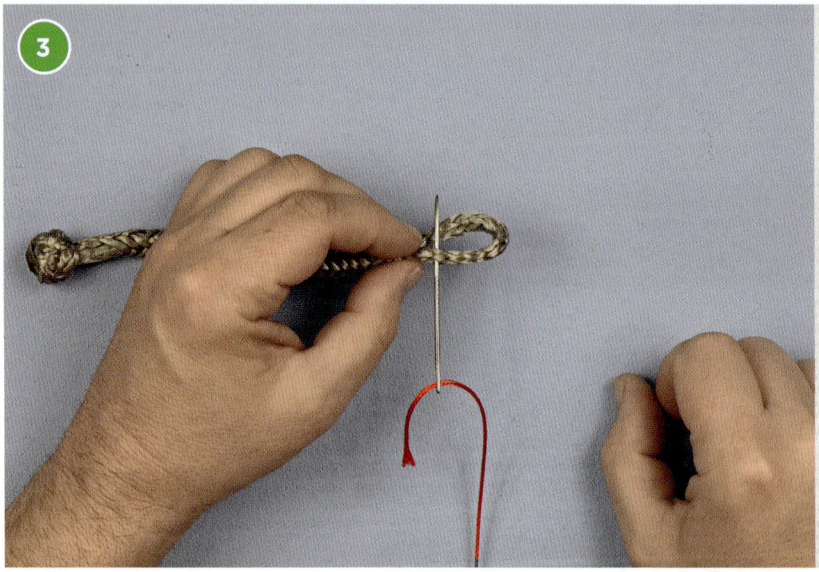

This gives the spot where the draw line is attached. Pierce through the rope by using a sailmaker's needle and some whipping twine or small-diameter string. You may also use 1 mm Dyneema, but polyester is equally suited.

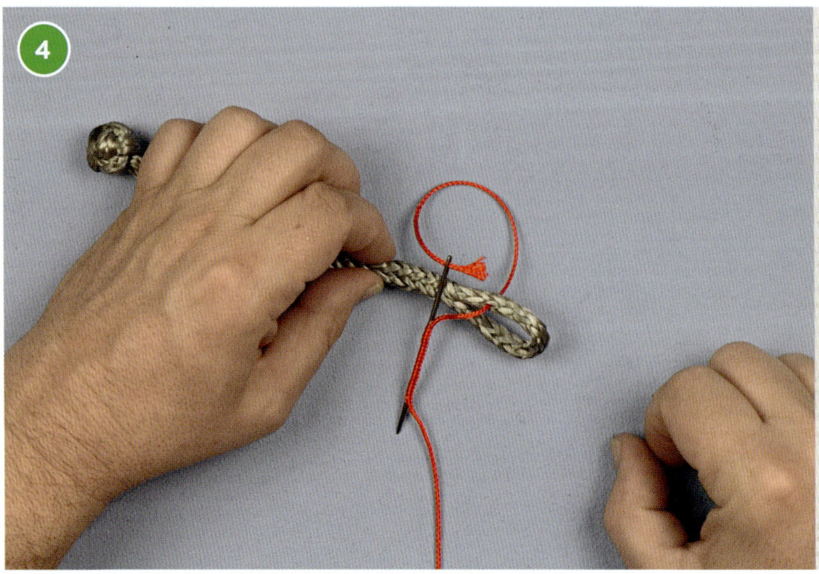

You may be able to splice the twine back inside itself over a short distance, using the sailmaker's needle.

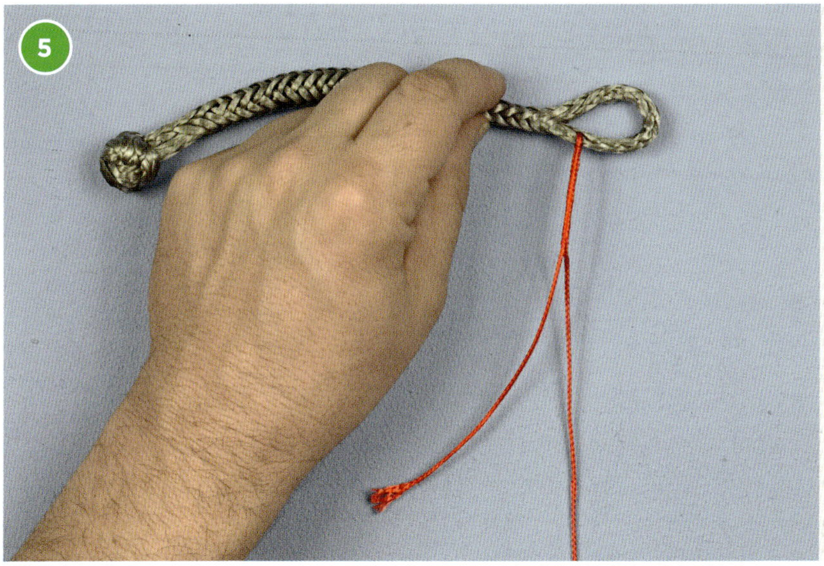

Pull tightly on the short end, and the draw line is neatly attached to the rope. Cut the long end and make a simple overhand knot so it is easier to grab the draw line.

Different versions of a draw line:

Left: The "quite all right" version. This draw line is not spliced back inside itself and forms an open loop.

Center: This version could be called "crafty with reason." The draw line is spliced and has a simple overhand knot at its end.

Right: The version "enamored down to the last tiny detail." The draw line is spliced, and its end is decorated by a perfectly round miniature diamond knot.

CAPTIVE SOFT SHACKLES METHOD 1

VIDEO 1.5.1

These captive soft shackles are made with two diamond knots in a row. Held by a small eye splice between the two diamond knots, it is impossible for the soft shackle to fall out or get lost when the shackle is opened. You can either make the soft shackle with both diamond knots and then splice a sheet or halyard around the section between the two diamond knots, or make the shackle with only one diamond knot, threading the rope's ends through an already finished eye splice in a sheet or halyard, and only then make the second diamond knot to the shackle. Just try to avoid finishing both the eye splice and the captive soft shackle by themselves, because then it will be too late to combine them, as can be seen to the right in the picture.

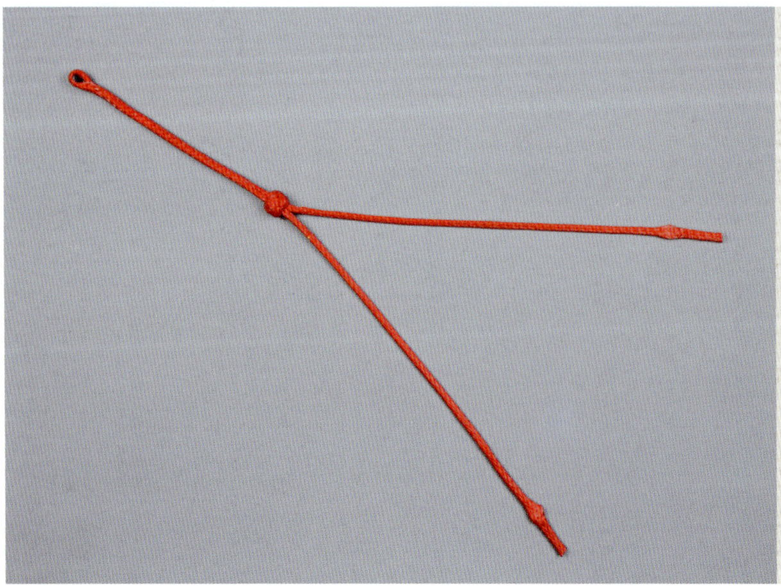

Tie one diamond knot and tighten it firmly. Lead the two rope ends, which need to be left longer, through a small eye splice at the end of a sheet or halyard, and then make the second diamond knot. In order to tie two diamond knots in a row, the ends of ropes up to 6 mm diameter need to have a length of about 50 cm (instead of 30 cm).

CAPTIVE SOFT SHACKLES METHOD 2

VIDEO 1.5.1

Position the diamond knot about 2–3 cm / 1 inch away from the end of the spliced section, so that you can tuck the shackle's little eye splice between the two parallel parts. This allows you to loop the soft shackle to the eye splice of a sheet or halyard. The connection between shackle and rope may not be 100% secure when the shackle is opened, but has proven to be highly reliable.

To close the soft shackle, place the small eye over the diamond knot as usual and pull it closed.

CLOSING A SOFT SHACKLE AND MAXIMIZING BREAKING STRENGTH VIDEO 1.5.1

Leave a small distance between the diamond knot and the end of the spliced section. Tuck the shackle's small eye splice through at this point and …

… place it over the diamond knot. By closing the soft shackle this way, the breaking strength is slightly increased, although this may hardly be necessary in practical use.

SIMPLIFIED SOFT SHACKLE VIDEO 1.5.1

Soft shackles can also be made in a simplified way by merely tucking the end through instead of splicing it back. This version looks less elegant. With plenty of benevolence, it can be argued that this method has the advantage that the soft shackle closes under load by itself more easily and that the load is distributed over the slightly greater surface of the two parallel-lying sections. In practical use there will be no great difference, let alone advantage. Working with a splicing needle, it even takes less time to fabricate a regular soft shackle than a shackle in the simplified, tucked version.

1

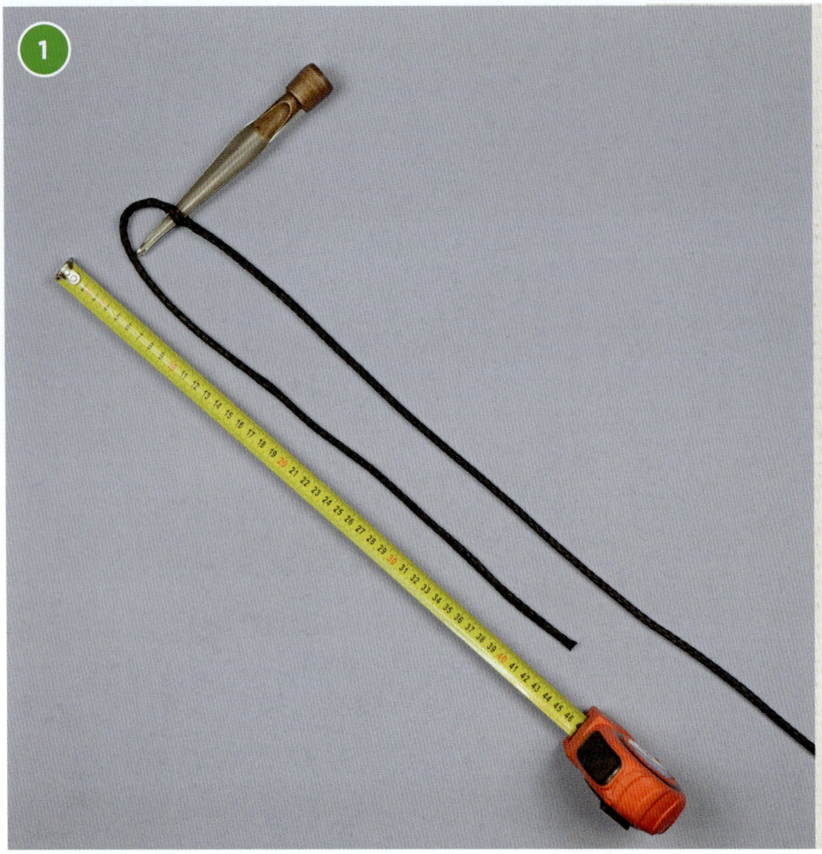

From the end of the rope, measure the desired circumference of the finished shackle (here, 15 cm), plus about 30 cm for the diamond knot. At this point, form a bight and pierce through the middle of the rope, using a marlinspike or, better, a Swedish fid.

Tip: *A length of 30 cm / 1 ft. is needed to tie and tension the diamond knot at a later stage and is sufficient for ropes up to 6 mm diameter. With ropes of larger diameter, this measurement needs to be increased. As a rule, the measured distance should correspond to at least 50 times the rope's diameter.*

2

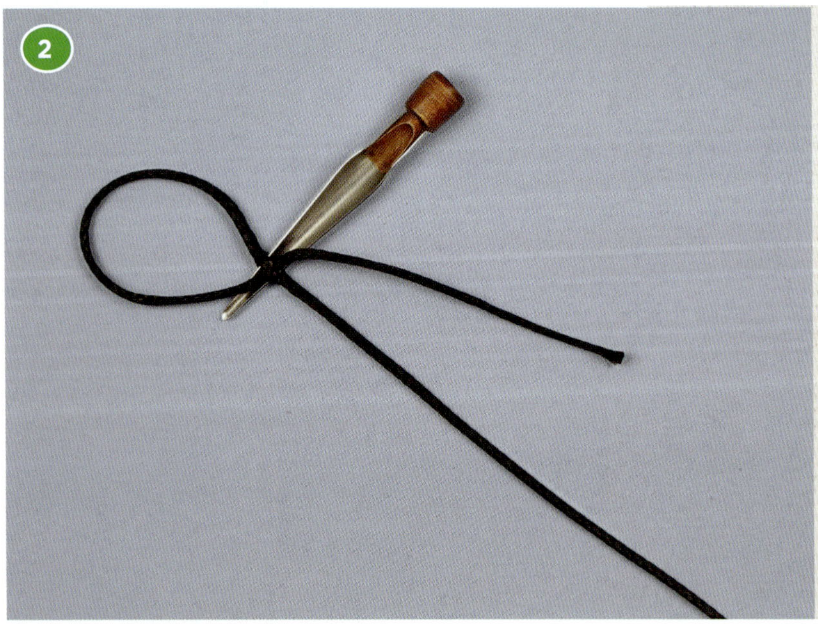

Tuck the end of the rope through.

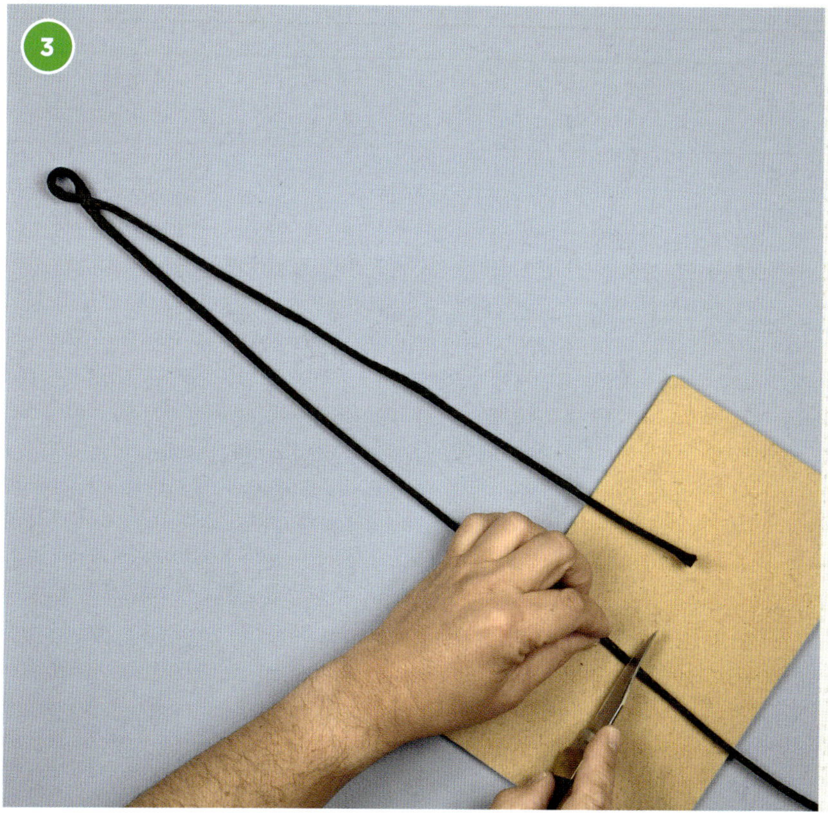

Pull the end through until a very small eye is formed with an inside diameter only slightly greater than the rope's diameter. Lead the standing end back parallel to the short end and cut it off to about equal length.

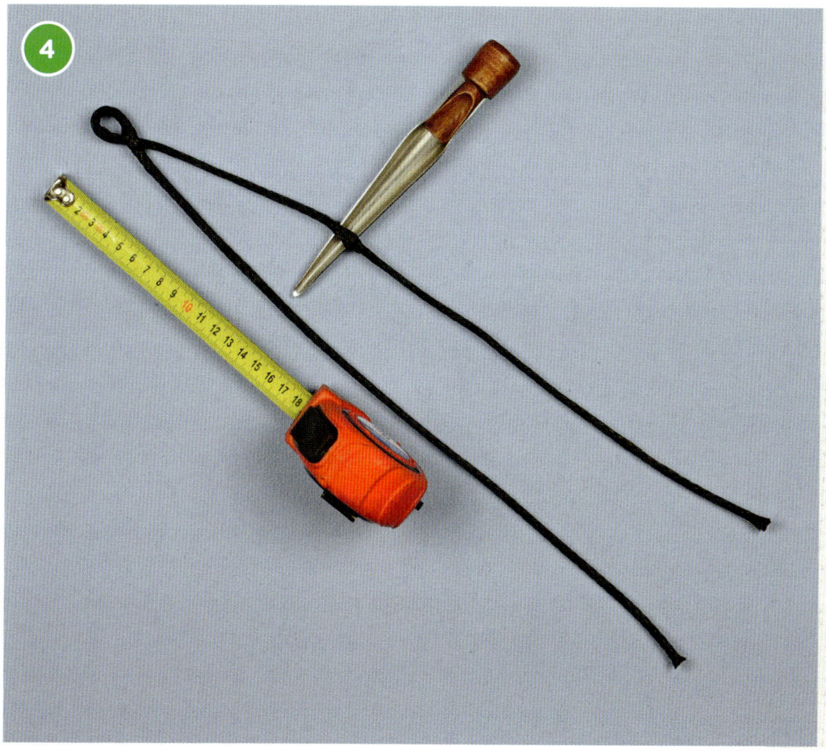

Measure the desired circumference of the finished soft shackle from the eye. At this point a brummel splice may be made to lock the ends before making the diamond knot. Using a marlinspike or Swedish fid, pierce through the rope at this point to tuck the other end through.

With a little practice, the two opposing tucks of the brummel splice are not really needed. They serve only to keep the two ends at equal length before tying the diamond.

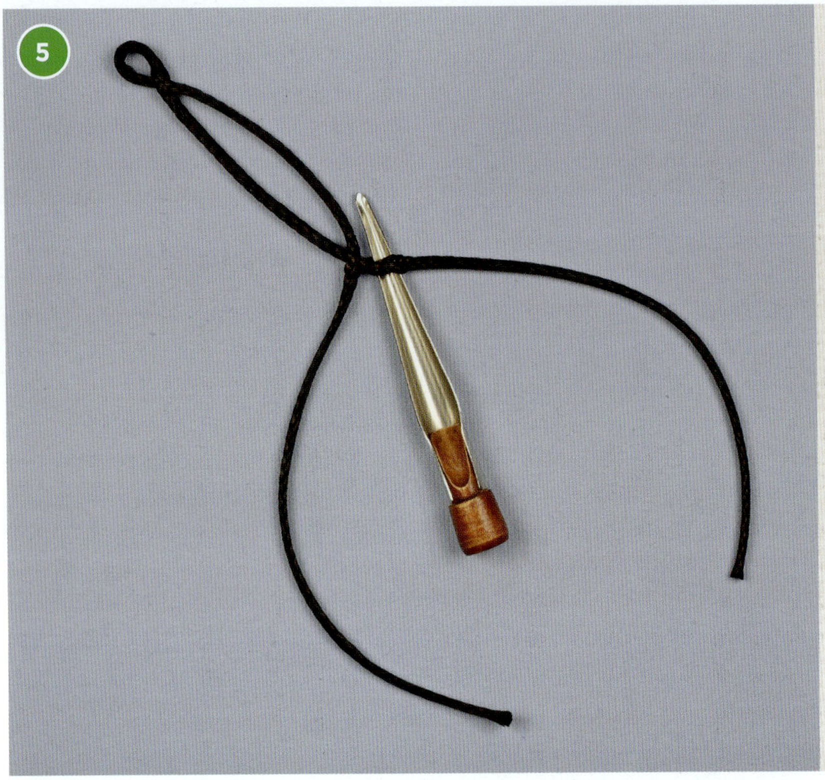

One end has been tucked. Now pierce through this end right behind the first tuck and pull the other end through. This gives the two opposing tucks of the brummel splice. Both ends should have the same length between the small eye and the brummel splice.

Important: It is not one end that is tucked back and forth through the other end, but, rather, one end is tucked through the other, and then this end is tucked through the first end!

Now tie the diamond knot with the two ends and tighten it firmly. Then cut the ends off and fuse them slightly—done.

SOFT SHACKLES MADE DIRECTLY WITH THE END OF A ROPE

It is also possible to fabricate a soft shacke directly out of the end of a single-braided Dyneema rope or out of the Dyneema core of a double-braided rope. The Dyneema core of a double-braided rope should be used only if the cover is meant to be stripped off for tapering the rope and if the core is suitably coated. Depicted on the left side in the picture is a soft shackle made directly out of the end of a single-braided Dyneema rope, and on the right side is a soft shackle made directly out of the Dyneema core of a double-braided rope.

SINGLE-BRAIDED HMPE ROPE (DYNEEMA)

VIDEO 1.5.2

1

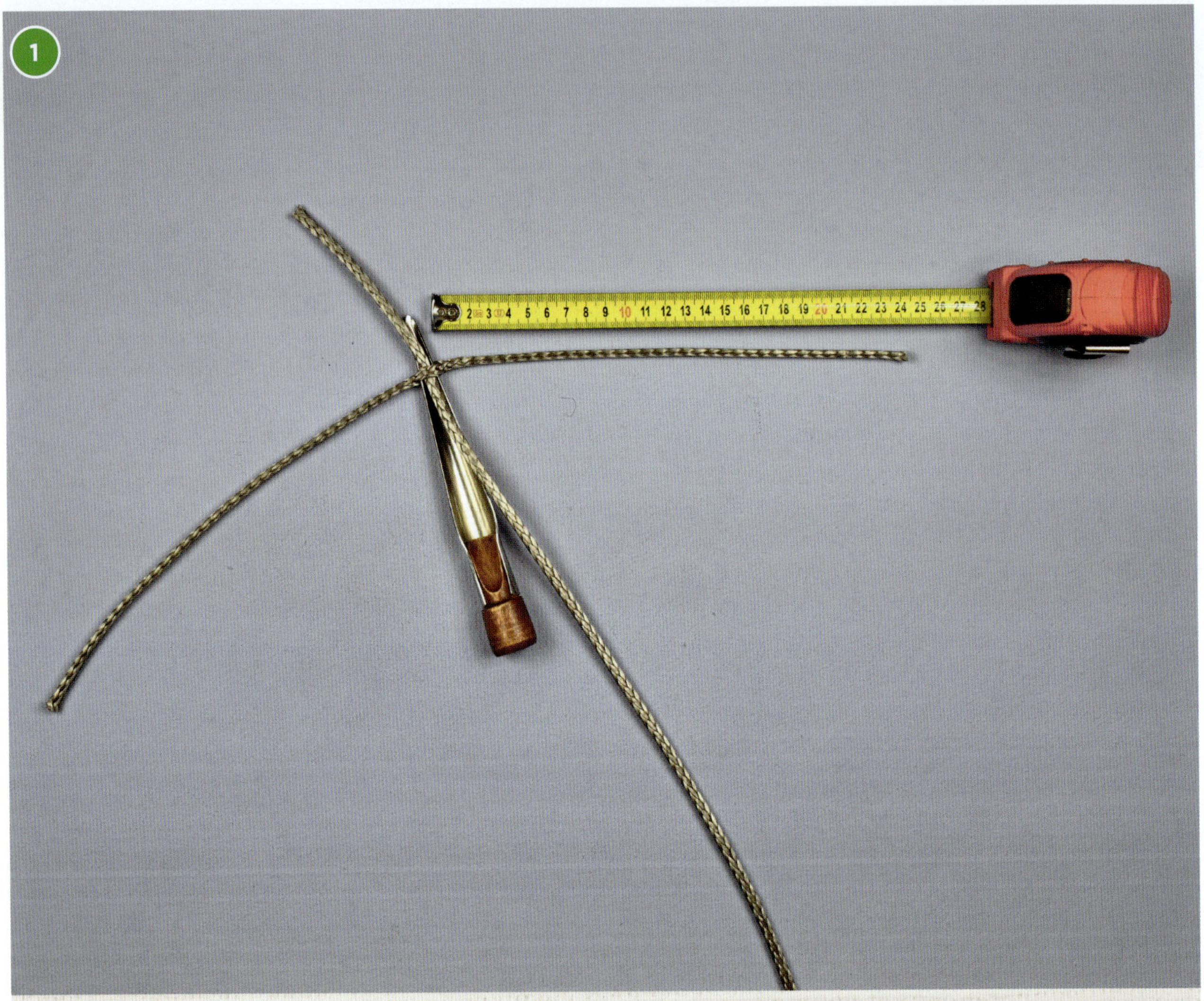

In order to fabricate a soft shackle out of the end of a single-braided rope, here 5 mm Dyneema, it is necessary to cut a short length off and splice this separate short length to the rope. Tuck the end of the rope through this separate piece, leaving a length of 48 times the diameter to one side (here, 24 cm to the right side) and a suitable length to tie a diamond knot (50 times the diameter) to the other side (here, 25 cm to the left side).

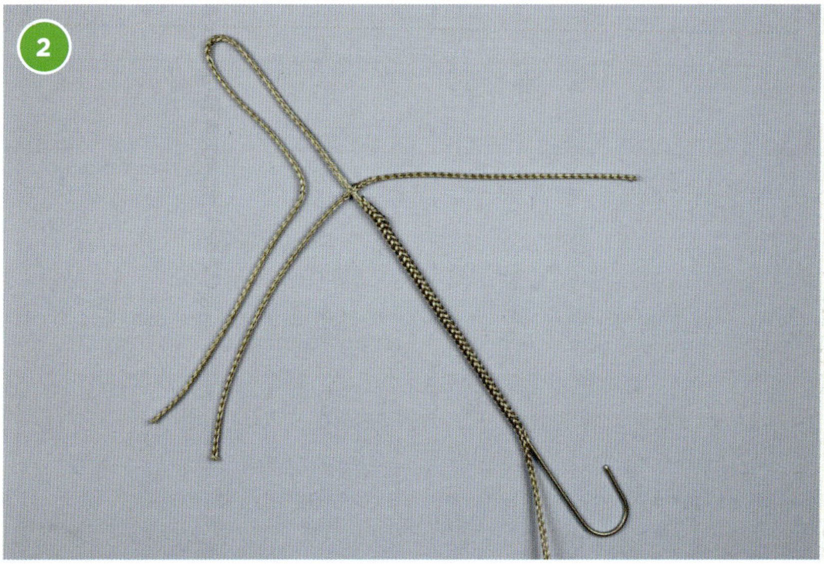

Pull the end of the rope through the separate piece of rope and lay out the desired circumference of the soft shackle (top). Lead the rope back and place the end parallel to the separate piece on the side with which the diamond knot will be made (left). Insert the splicing needle as shown and exit close to where the separate piece runs across.

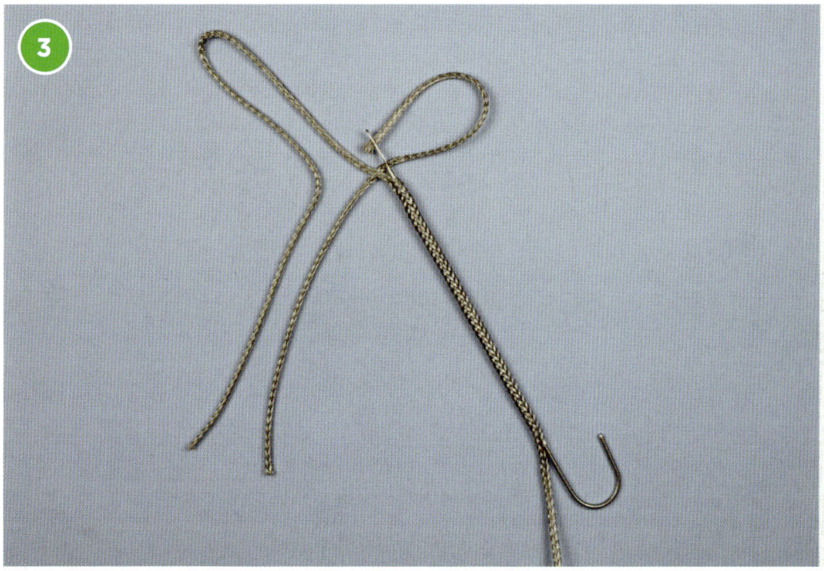

Now the right end of the separate piece is spliced into the rope. Thread the end onto the splicing needle and pull it inside the standing end. Taper the end over the last third of its length and let it disappear inside the standing end. All details about tapering can be found in the chapter "Basic Splicing Skills" on page 20.

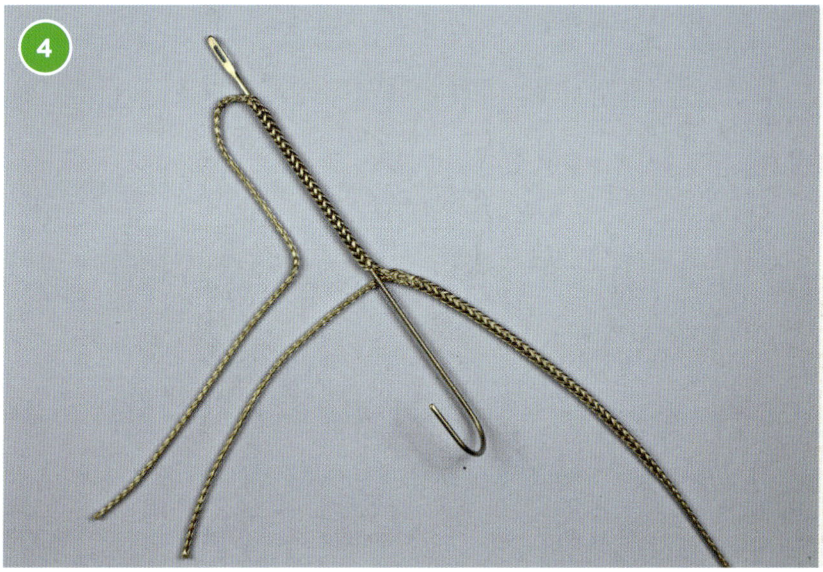

Insert the splicing needle right above where the short separate piece runs across, and push it through over the distance of the desired circumference of the finished soft shackle.

5

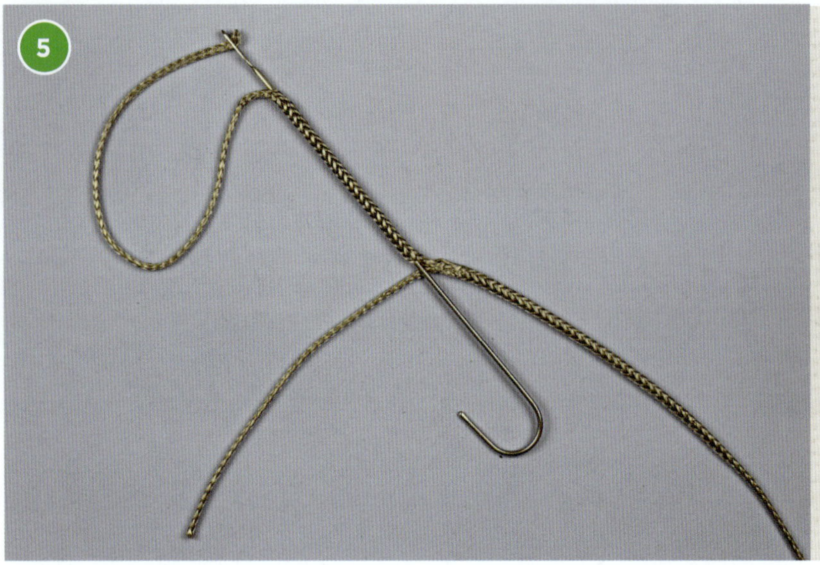

Now thread the end of the rope onto the splicing needle and pull it through.

6

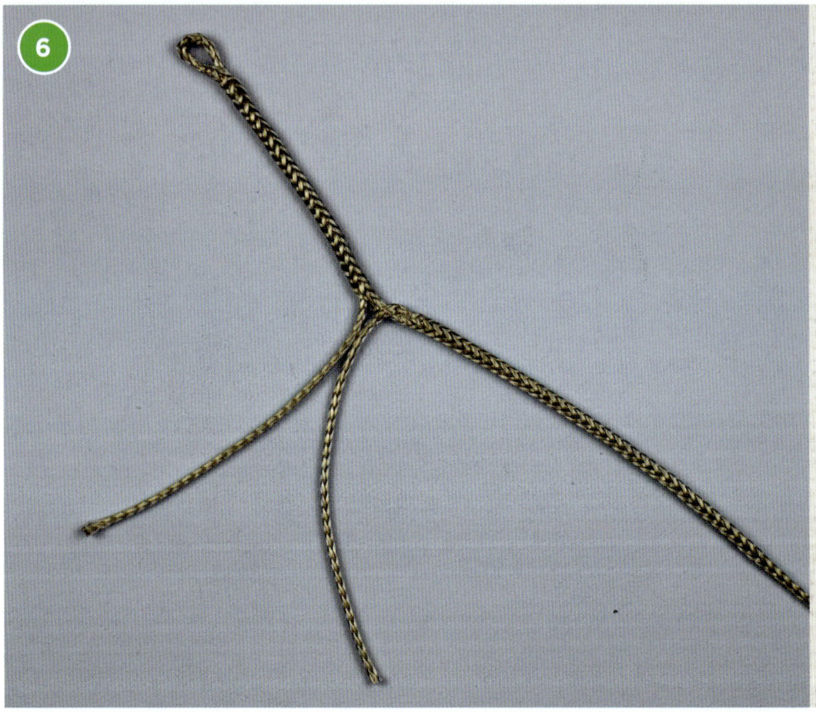

Pull the end through until the eye splice is very small. (To be exact, the inside diameter of the eye should be equal to the outside diameter of the spliced section right below the eye.) Now tie a diamond knot with the two protruding ends.

Optionally, the two ends can be first tucked back through the rope in order to hold them a little better in position before making the diamond knot, although this is not really necessary.

7

Add a draw line to the eye splice if desired.

Left: A soft shackle made directly out of the end of a single-braided rope —a very elegant solution

Right: A separate soft shackle held by an eye splice at the end of a single -braided rope. This arrangement yields slightly higher breaking strength and has the added advantage that rope and soft shackle can be separated if needed.

Double-Braided rope with HMPE core (Dyneema)

Video 1.5.2

First tie a loop knot, preferably a butterfly knot, about 3 meters from the end of the rope. Measure the following lengths from the end of the rope: 50 times the diameter of the core (for the diamond knot) plus twice the desired circumference of the soft shackle, and substract the length of the core shortening, which is about six times the rope's diameter. At this spot, extract the core.

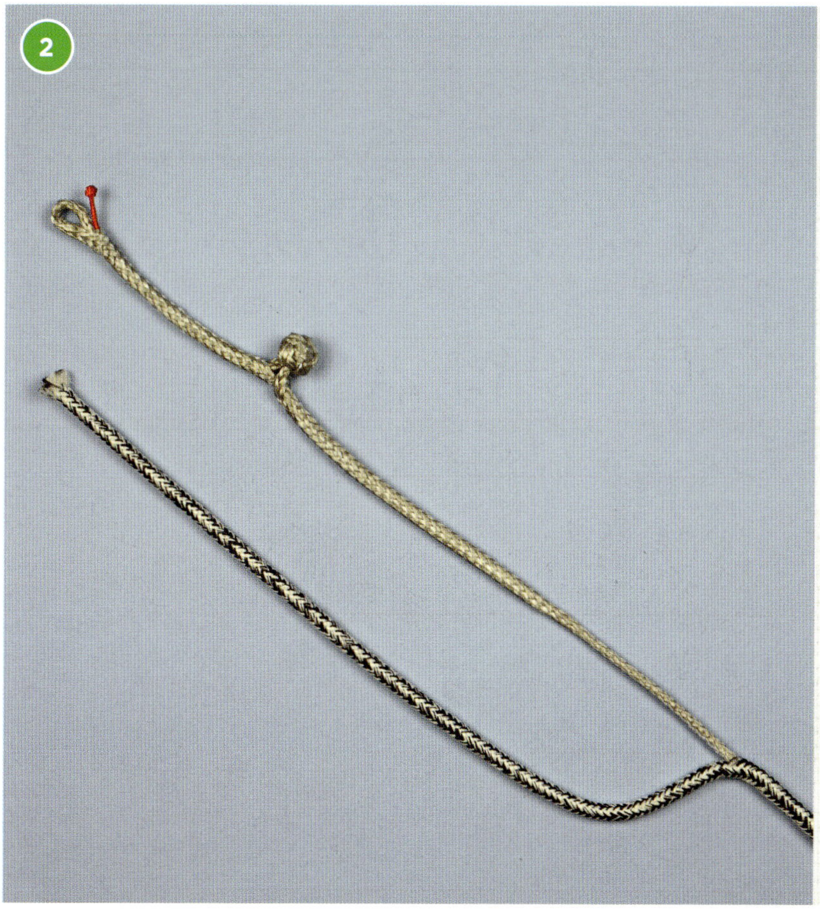

Pull the core out farther and make a soft shackle exactly the same way as with a single-braided rope (see above). You will again need a separate length of core, which is spliced to the rope.

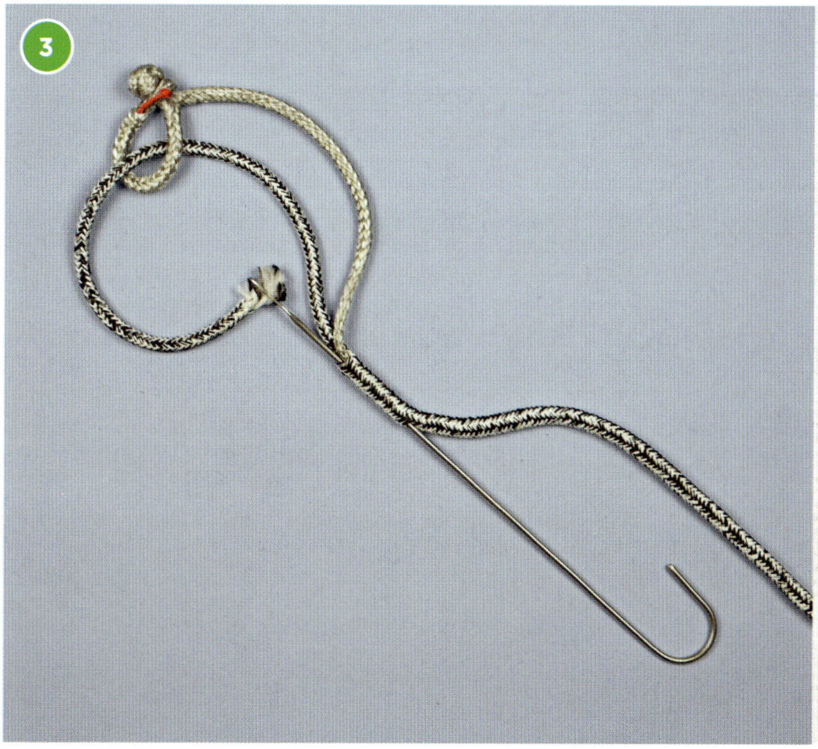

Close the soft shackle, tuck the cover through the shackle, and pull it a short distance back into the standing end, using the splicing needle. Important: Do not insert the splicing needle into the core that is inside the standing end. The splicing needle must be pushed through between core and cover!

Pull the cover end tight until it is snug around the soft shackle and next to the diamond knot. Cut the protruding cover end off, taper its end, and let it disappear inside the standing end.

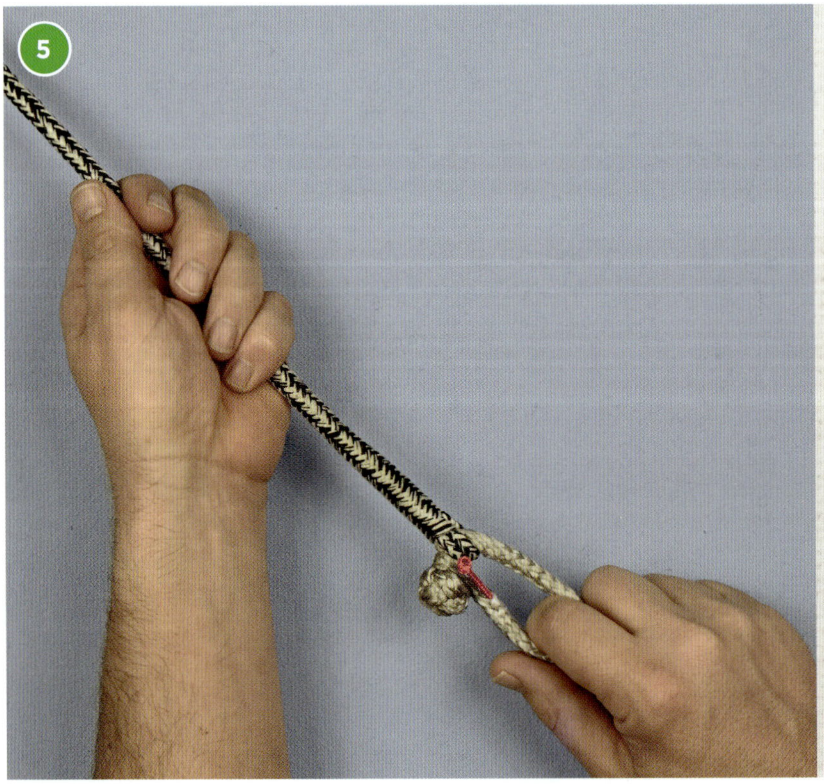

Hook the butterfly or loop knot to a strong point and milk back the cover until all of the core length disappears inside and the soft shackle is right at the end of the rope.

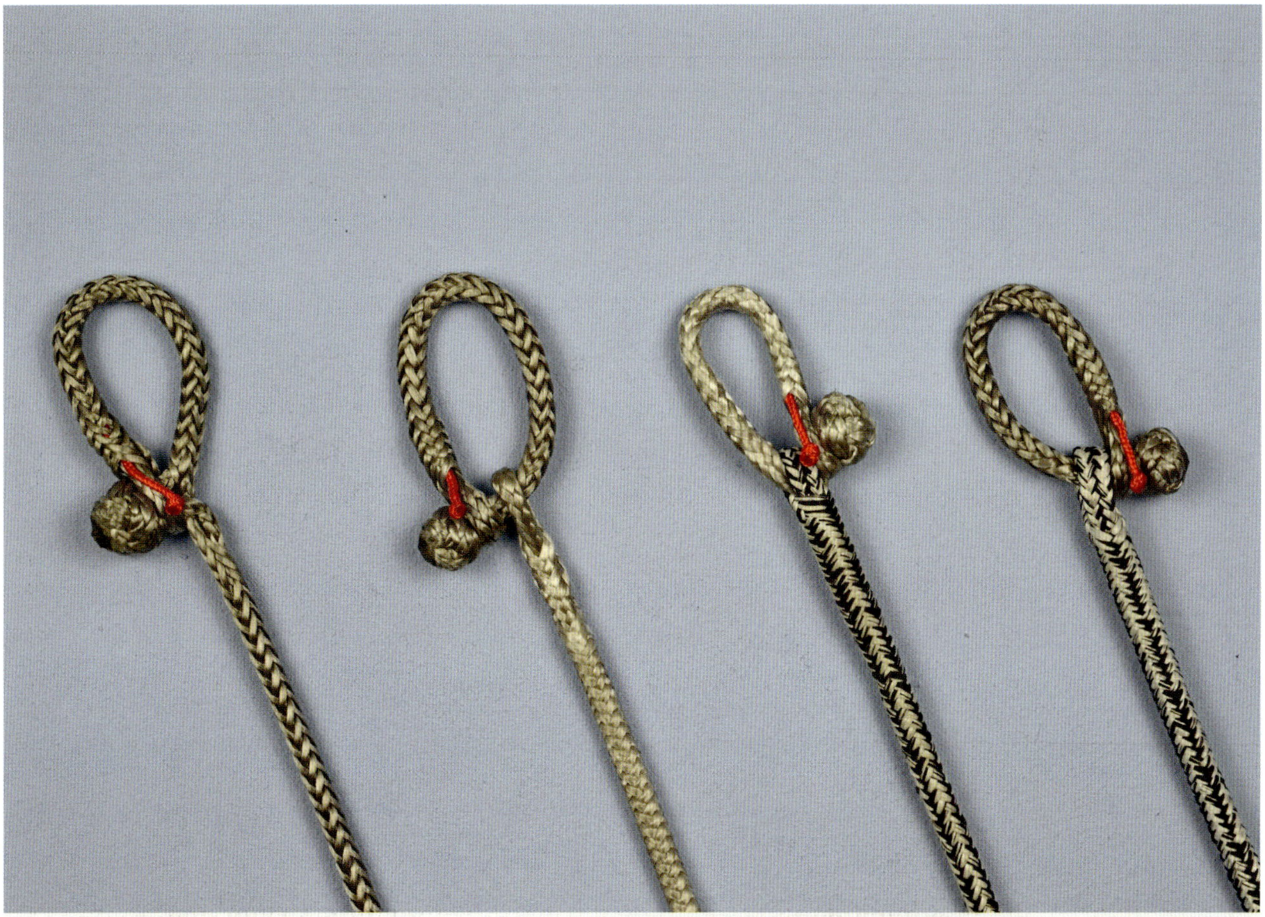

Different splicing solutions for ropes terminating with a soft shackle, from left to right:

- soft shackle made directly with the end of a single-braided rope (Dyneema)
- separate soft shackle held by an eye splice of a single-braided rope (Dyneema)
- soft shackle made directly out of the core (Dyneema) of a double-braided rope
- separate soft shackle held by an eye splice of a double-braided rope

Fabricating a soft shackle directly out of the end of a rope is a very elegant solution, but there are draw backs in technical as well as in practical terms. Compared to a soft shackle that is simply held by a small eye splice at the end of a sheet or halyard, the breaking strength is slightly diminished, and it is not possible to disconnect the soft shackle from the rope. In practical life, the soft shackle wears faster than the rope itself and needs to be replaced at an earlier stage.

There is hardly any limit in size for a soft shackle. Here a giant soft shackle made out of 60 mm Dyneema …

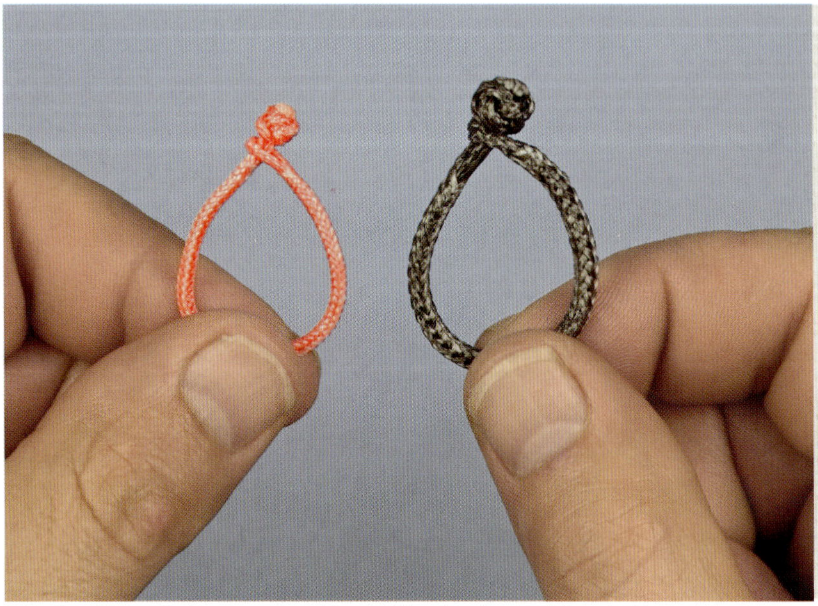

… and tiny soft shackles, which are nevertheless amazingly strong from 1 mm and 2 mm Dyneema, respectively.

1.6 MODERN LASHING "TRANSFILAGE" MADE WITH DYNEEMA, SPECTRA

With single-braided high-modulus fiber ropes such as Dyneema, Spectra, or Vectran, it is possible to fabricate textile connections, so called lashings, which are extremely strong if tied correctly.

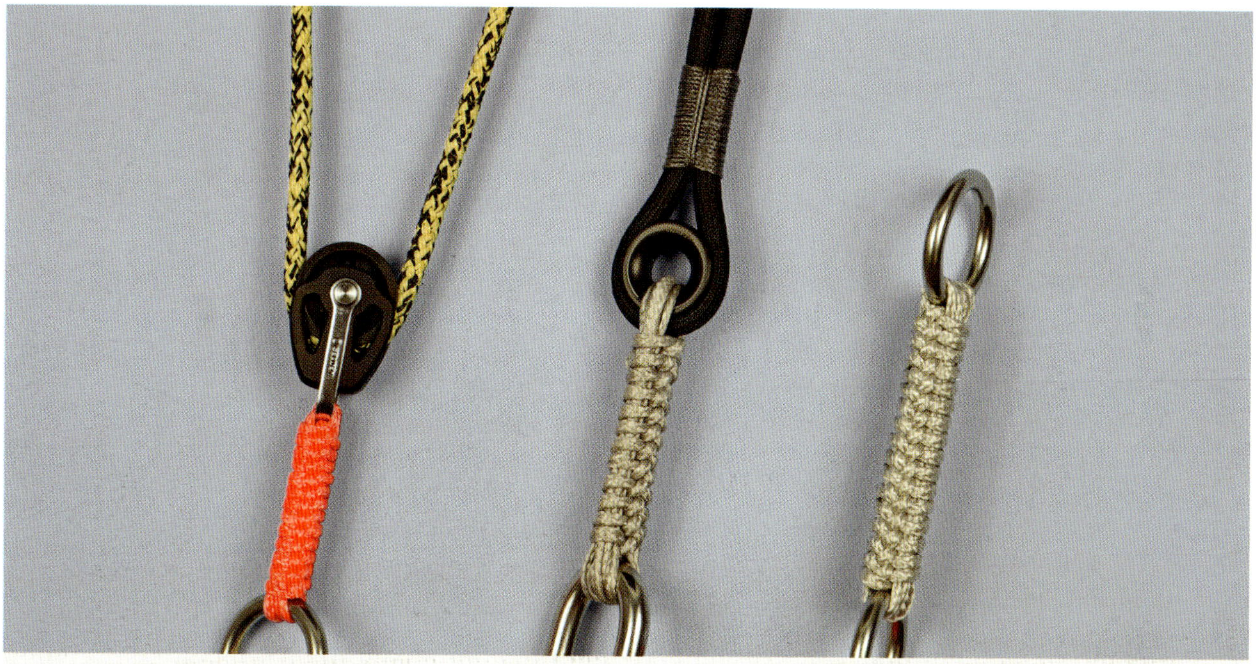

For long-term attachments, the modern lashing out of high-modulus fiber rope offers a very good solution. Many such modern rigging solutions originated from professional offshore sailing in France, where they are called "transfilage," a term that is often used universally. Knots generally greatly diminish breaking strength in single-braided, high-modulus fiber ropes, which is why tying a lashing in the correct way is of utmost importance. A lashing should have a maximum of four continous turns around the attachment points, which give a total of eight parts of the rope between those attachment points, four on the front and four on the backside of the lashing. With more than four turns, the load would no longer be distributed equally between all parts of the rope. The lashing is tied off by a series of half hitches. In principle, different patterns of half hitches could be used, such as in ringbolt hitching, but shown here is solely the pattern that became common practice in modern lashings. It will only rarely be necessary to use rope of more than 4 mm diameter for a lashing, because that alone yields breaking strengths of up to 6 metric tons (13,228 lbs.). A good length of single-braided Dyneema or Spectra should be carried on board every sailing yacht because it allows for countless repairs and fixtures to be improvised; even broken turnbuckles can be replaced or secured. Typical applications of lashings include attaching blocks to fittings or guardrails, attaching life lines to stanchions, or even connecting composite shrouds of modern offshore trimarans to the chain plates. Compared to soft shackles, a lashing has the advantage that it can be applied under tension.

START

VIDEO 1.6

Construction:	single braided (no core)
Material:	high-modulus fiber (e.g., Dyneema, Spectra)
Difficulty:	easy
Splicing tools:	splicing needle or fid

1

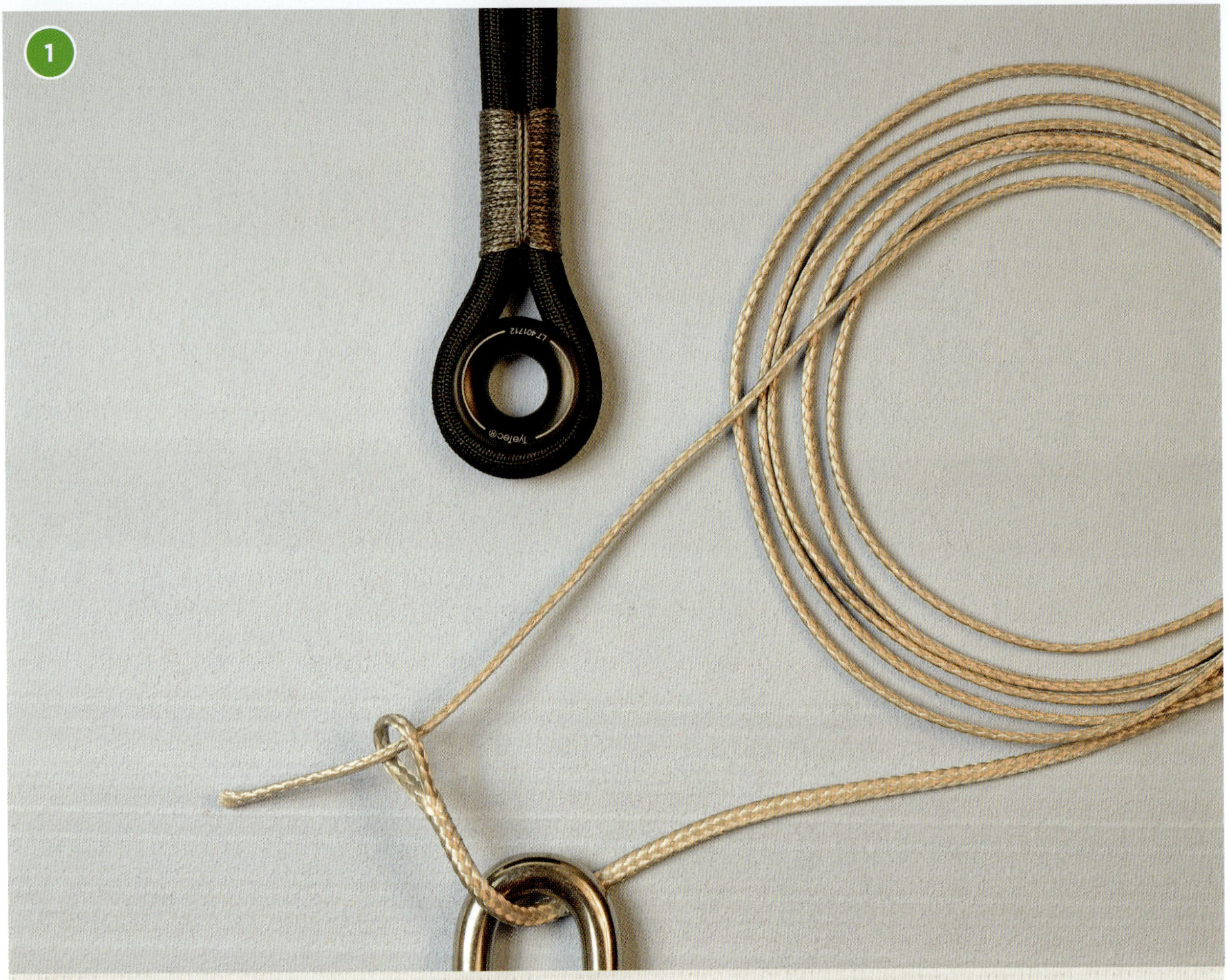

Make a small eye splice at the end of the rope (see chapter 1.2) and loop the other end through, as shown, to connect to one of the attachment points. If an eye splice is looped to an attachment point, the splice does not need to be secured by a brummel splice or a whipping.

The length of rope needed for a lashing is made up by the number of turns around the attachment point, a maximum of four turns giving eight times the distance between the attachment points plus the length over which the lashing is to be covered by half hitches; calculate about 1 meter for 6 cm of covering.

Wrap the rope in continous turns around or through the attachment points. Make no more than four turns around, so that a maximum of eight parts of the rope lead from one attachment point to the other.

It is best to already apply tension at this stage, so that the turns can straighten out and bed down in their final position.

After leading the end around the attachment point for the last time, tie a half hitch around the strand that is farther away (here, the strand at the front).

Then tie a half hitch around the other strand, as shown (here, the one at the back). Keep tying half hitches, always alternating between going around the front and the back strand.

Here is a view from the side: tie a half hitch around the strand to the left and ...

... then lead the rope across and tie a half hitch around the strand to the right. Make a minimum of four half hitches around each strand. It is not necessary to cover the entire length of the lashing with half hitches, but it looks good, protects against chafe, and adds stiffness when the lashing is not under tension.

To finish off, tie an overhand knot to serve as a stopper. The overhand knot is placed at the back side of the lashing.

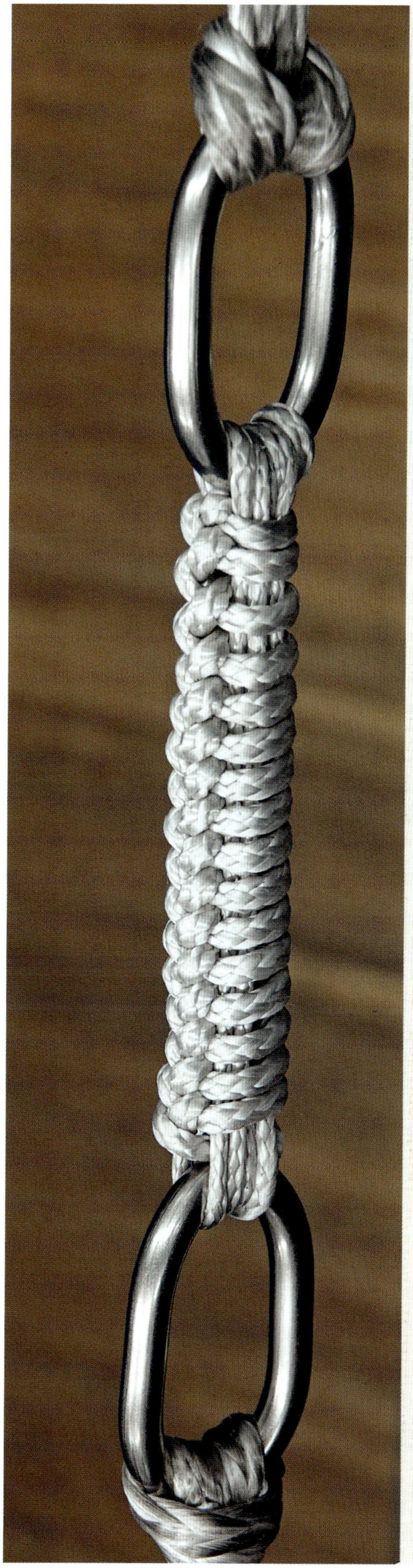

A lashing from 4 mm Dyneema tensioned to 2.5 metric tons (5511 lbs.). The high-strength round rings from AISI 316 stainless steel have already deformed considerably. Breaking strength of this kind of lashing is a minimum of 4 metric tons (8,818 lbs.).

SPLICING DOUBLE-BRAIDED ROPES (ROPES WITH CORE AND COVER) 2.1 EYE/THIMBLE SPLICE IN CONVENTIONAL FIBER ROPES

Double-braided ropes with core and cover from polyester (PES) or polyamide (PA) or polypropylene (PP).

Conventional fiber ropes from polyester are hard wearing and long lasting. They are a good choice for cruising yachts. On board a more performance-oriented boat, polyester is used only where stretch is not a problem, and maximum strength is not of prime importance. Ropes from polyamide (nylon) have even more stretch and are more suitable for mooring lines and anchor rodes. Ropes from polypropylene are lower priced and have the advantage that they float in water.

Splicing the end of the mainsheet to a block of the mainsheet tackle or purchase looks much better and more seamanlike than attaching the sheet with a thick knot. In a four-part tackle, the load is greatly distributed onto four times the section of the rope. Also, the maximum length of the mainsheet is comparably a lot shorter than is the case with a halyard or foresail sheet, so that stretch characteristics are of lesser importance, and in many cases polyester ropes are perfectly fine for mainsheet tackles.

START

VIDEO 2.1

Construction:	double braided (core and cover, kernmantle rope)
Material:	core: polyester, cover: polyester
Difficulty:	medium
Splicing tools:	splicing needle or similar pull-through tool necessary; fids are suitable only for loosely braided ropes.

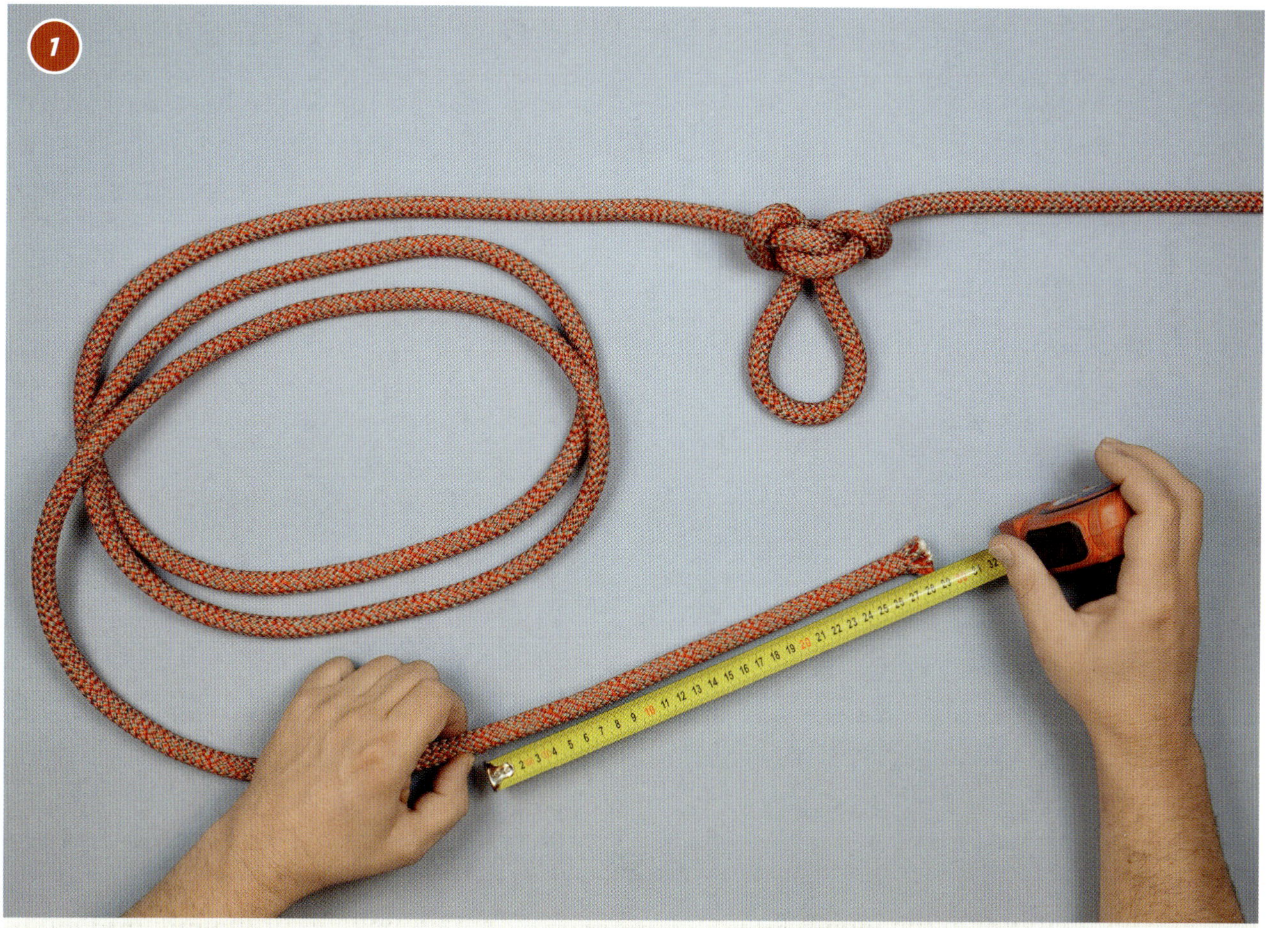

Tie a loop knot, preferably a butterfly knot, about 3 meters / 10 ft. from the end of the rope. Cut the end of the rope with a knife; it must not be fused. Measure the splicing length of the cover. This length is 24 times the diameter of the rope. Here, 24 x 12 mm diameter = 288 mm or, rounded off, 29 cm.

At the measured distance, pull out a single yarn to mark this spot. Use a metal needle or small spike to lift up a single yarn of the cover braid.

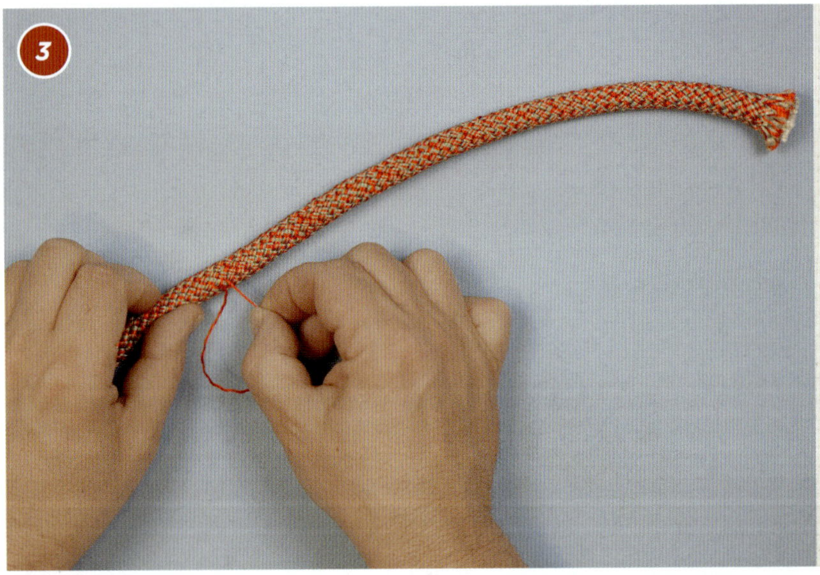

Pull only on the side of the yarn that leads toward the short end of the rope. Be careful not to tension the yarn toward the side of the standing end, since this would distort the cover braid.

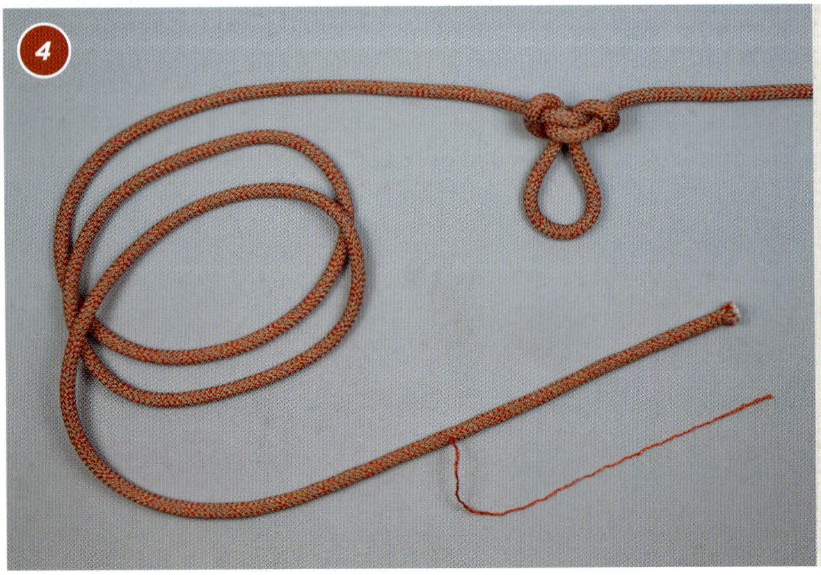

Pull the yarn out completely. This is now the cover marker yarn.

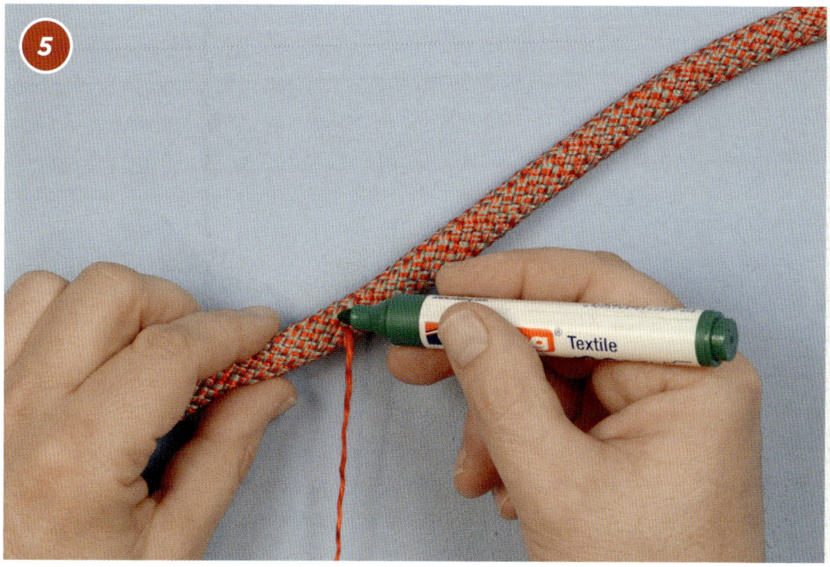

Not so good, but as an alternative it is also possible to simply mark the measured distance with a felt pen instead of pulling out a marker yarn.

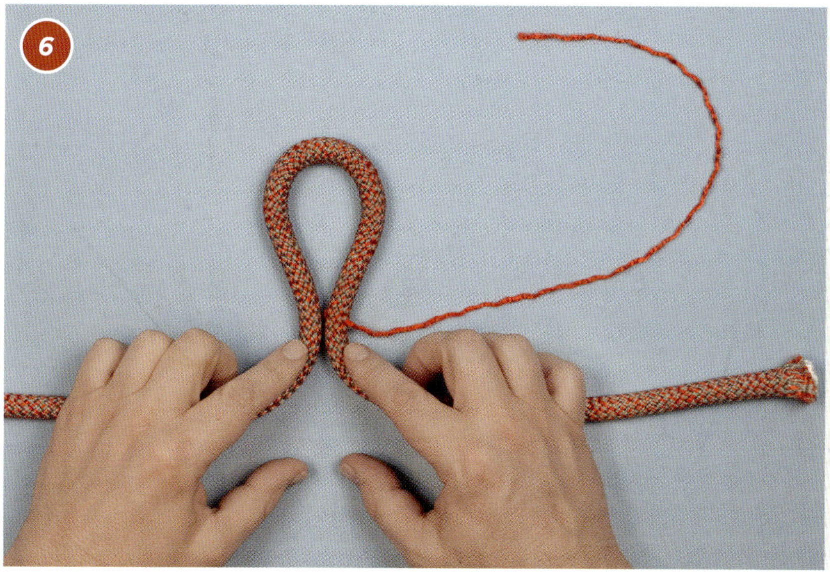

Now the size of the eye you want to splice has to be determined. The circumference of the eye starts at the cover marker yarn. Form the eye in the desired size.

If a thimble is to be spliced into the eye, lay the rope around the thimble to pick off the eye circumference needed.

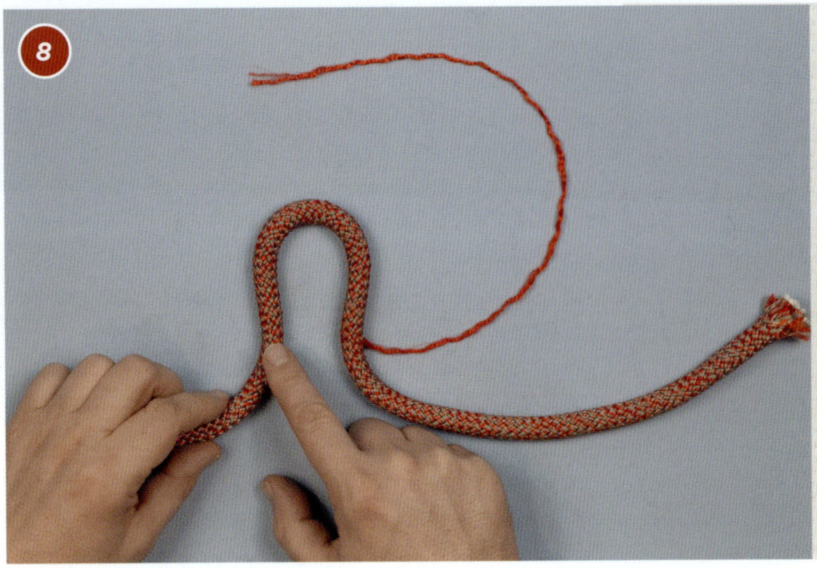

At exactly the end of the eye circumference lies the spot where the core has to be extracted.

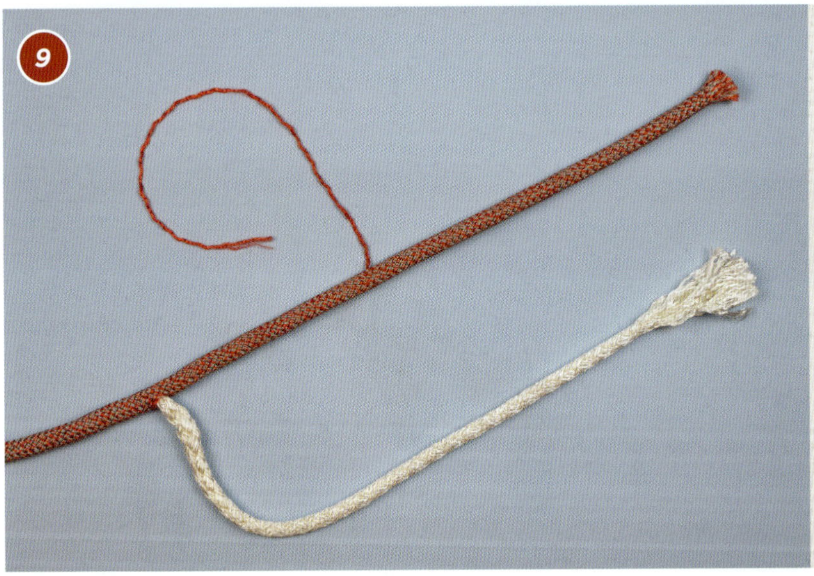

Extract the core. Everything about how to get the core out of the cover can be found in the chapter "Basic Splicing Skills" on page 20.

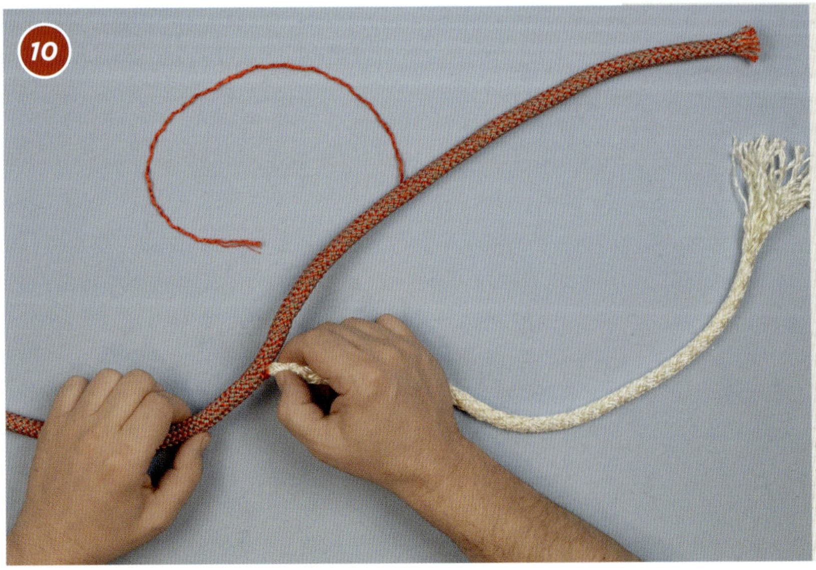

Hold the core at exactly the spot where it exits from the cover. It's best to grip it with your thumbnail. Do not let go.

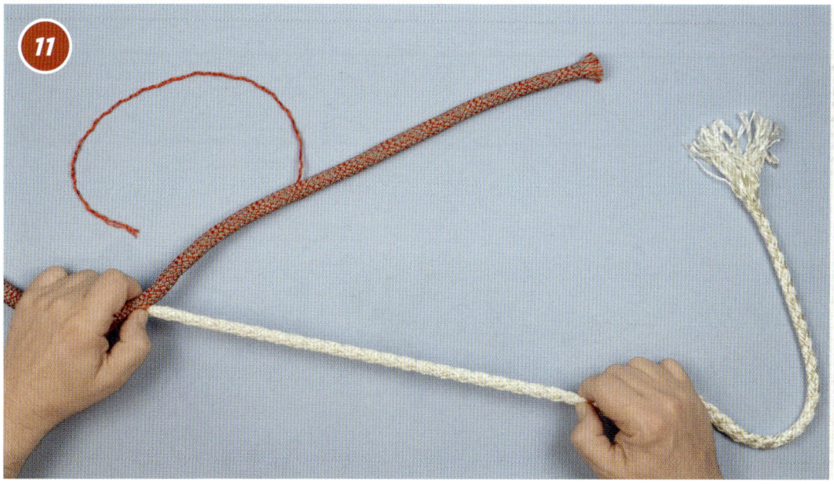

Pull out more of the core.

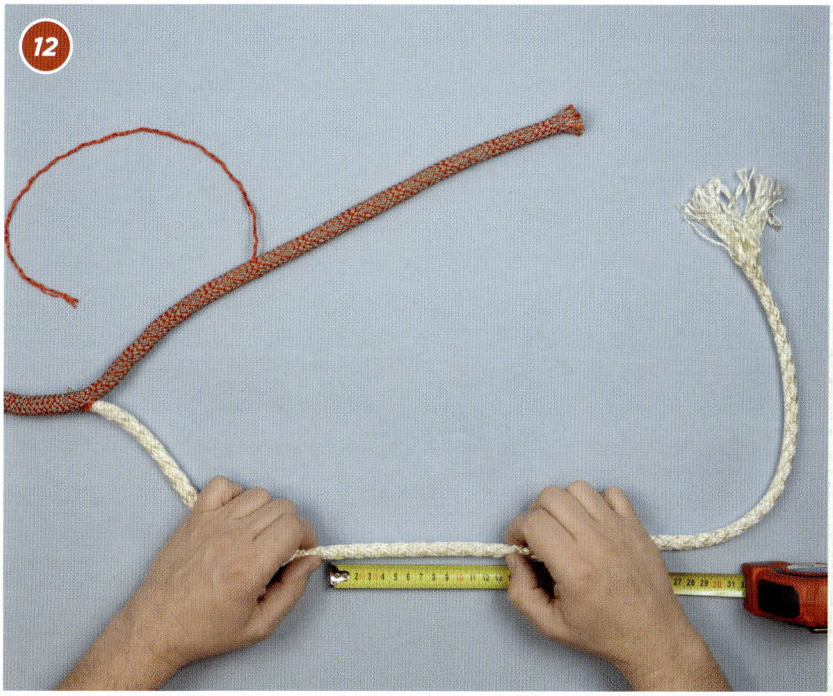

From the spot where you are still holding the core with your thumbnail, measure the so -called core shortening away from the end of the core. The length of the core shortening is 10 to 12 times the rope's diameter. For a 12 mm rope such as shown here this comes to 12–14.4 cm.

***TIP:** The exact length of the core shortening varies slightly between different rope types and manufacturers and depends on the type of braid and the ratio of volumes between core and cover.*

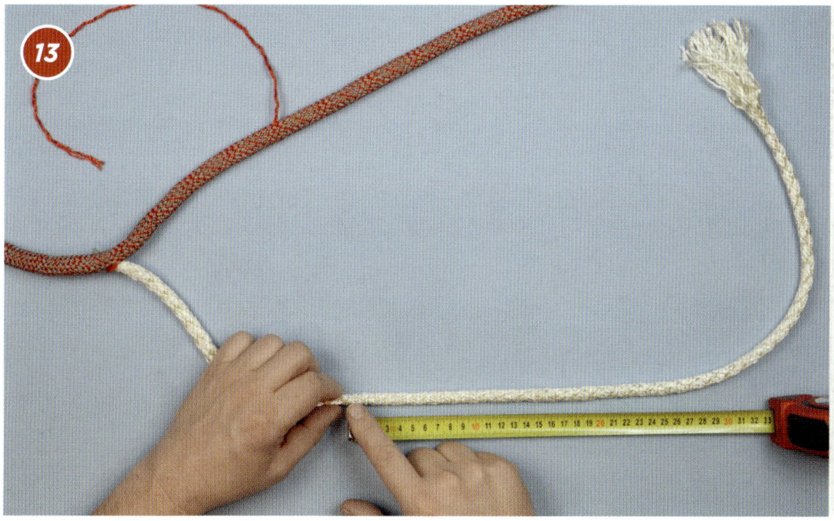

After you have measured the core shortening, this spot needs to be marked.

Again pull out a single yarn. Use a metal needle or small spike to lift one single yarn up.

Pull only at the side of the yarn that leads toward the short end of the core. Be careful not to tension the yarn toward the other side, so as not to distort the braid farther up the core. Pull the yarn out completely. This makes the core marker yarn.

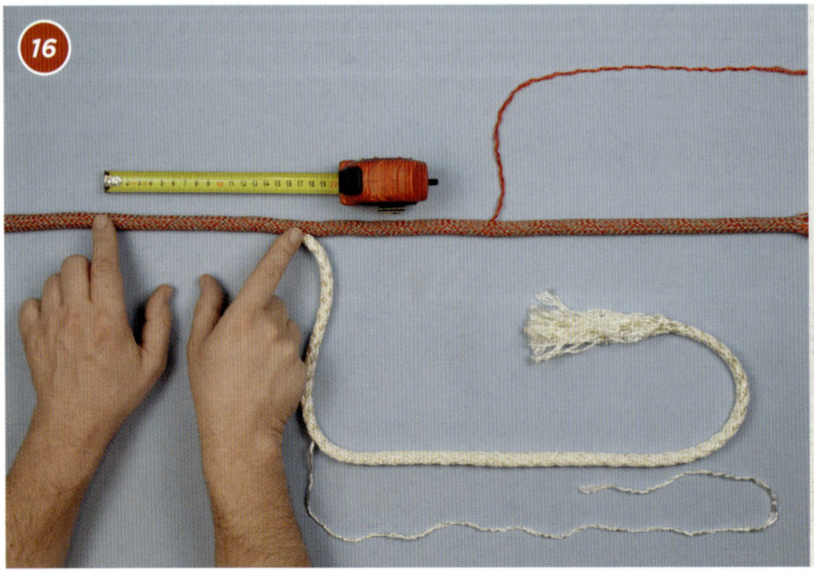

Now measure along the standing end, starting at the spot where the core exits. The length to measure is 14 times the rope diameter; here, 14 x 12 mm diameter = 168 mm or 17 cm rounded off. Smooth out the bunched cover braid before measuring. At the measured distance lies the spot where to insert the splicing needle.

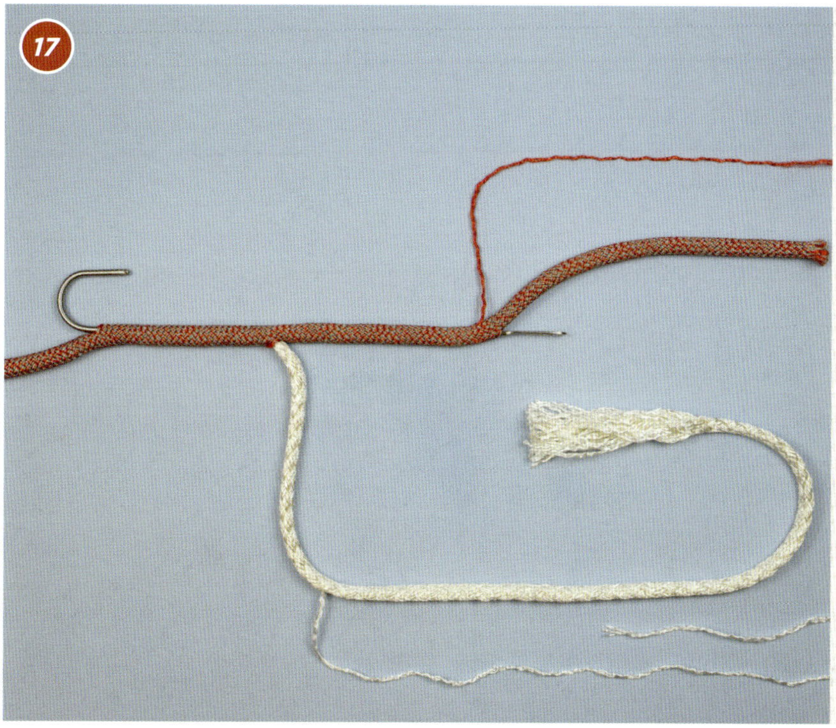

Push the splicing needle through as shown, going between the core and cover of the rope. Bring the tip of the splicing needle out one rope diameter behind the cover marker yarn. On what side around the rope should you insert the splicing needle? It's best opposite the side on which the core exits. And on what side around the rope should you bring out the tip of the splicing needle? On the same side on which the core exits. All details about working with the splicing needle can be found in the chapter "Basic Splicing Skills" on page 20.

It is of utmost importance not to pierce into the core when pushing through the splicing needle. Check if the splicing needle has gone through correctly by pulling out more length of the core. The core must easily slide along the splicing needle.

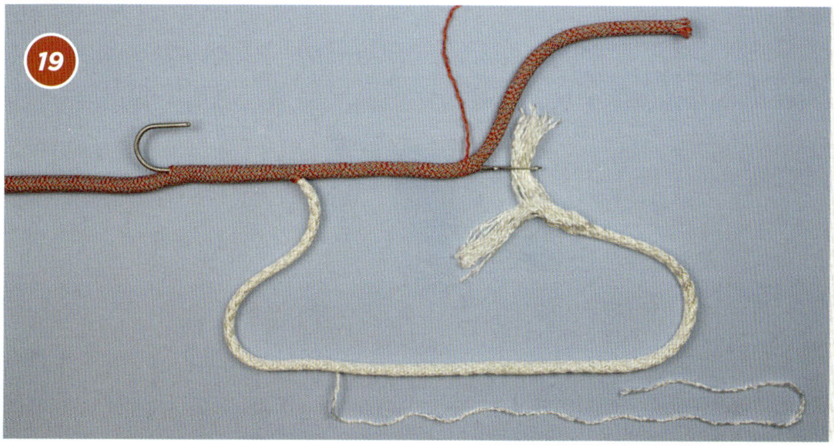

Untie the braid at the end of the core. Thread a little more than half of the core yarns onto the splicing needle, as shown.

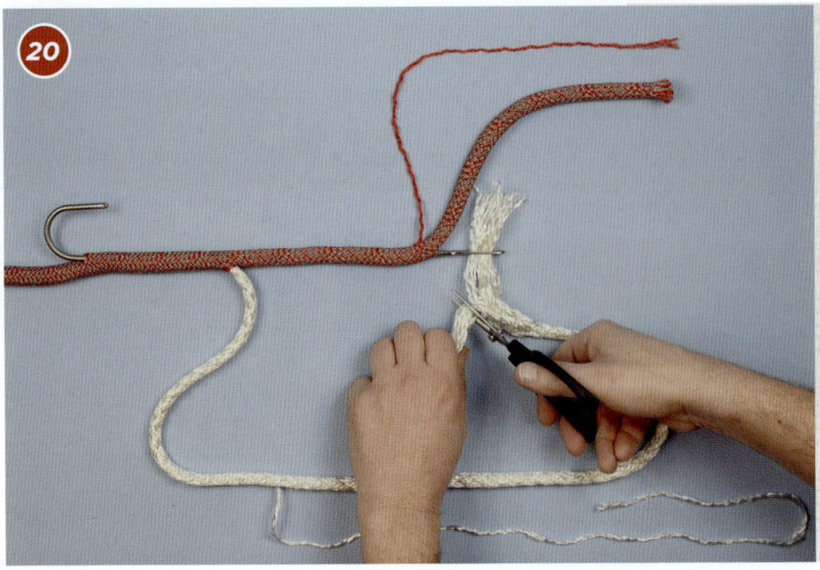

The remaining yarns are cut off carefully.

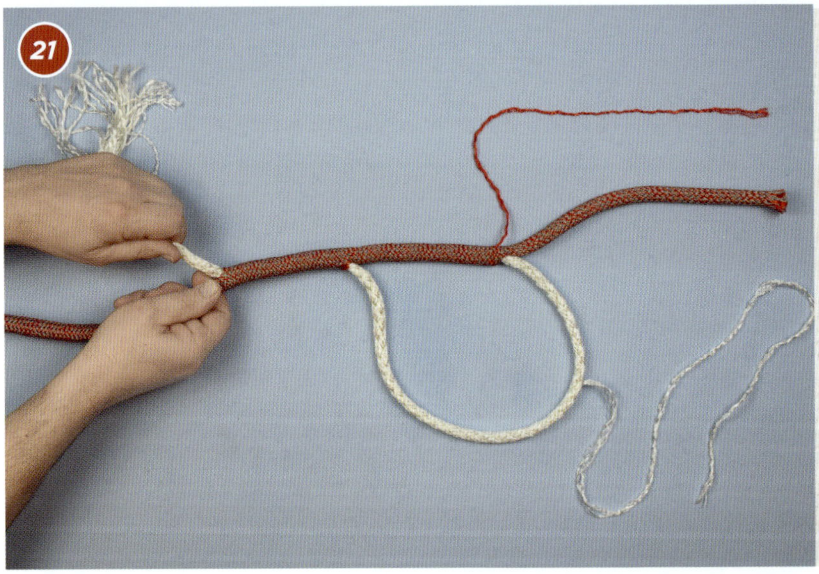

Pull the core through with the splicing needle and pull it farther by hand until …

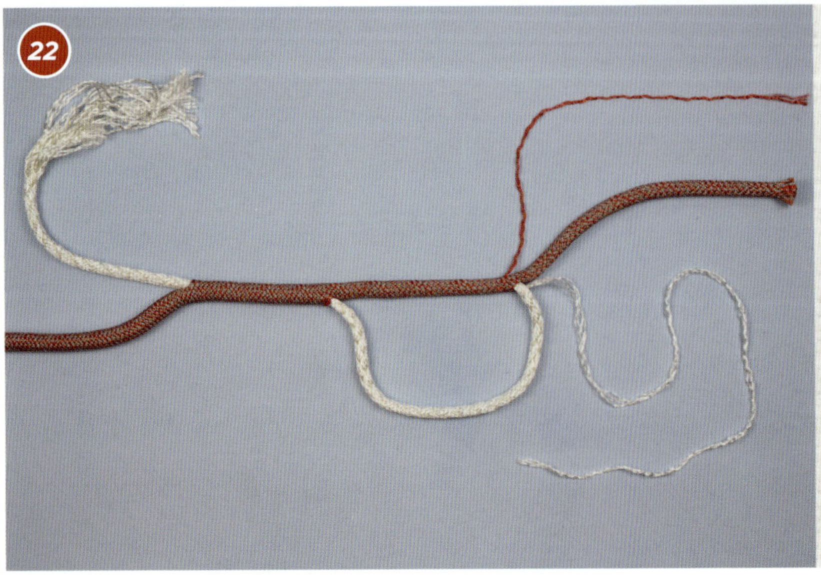

… the core marker yarn reaches the spot where the core enters inside the cover, as is the case in the picture.

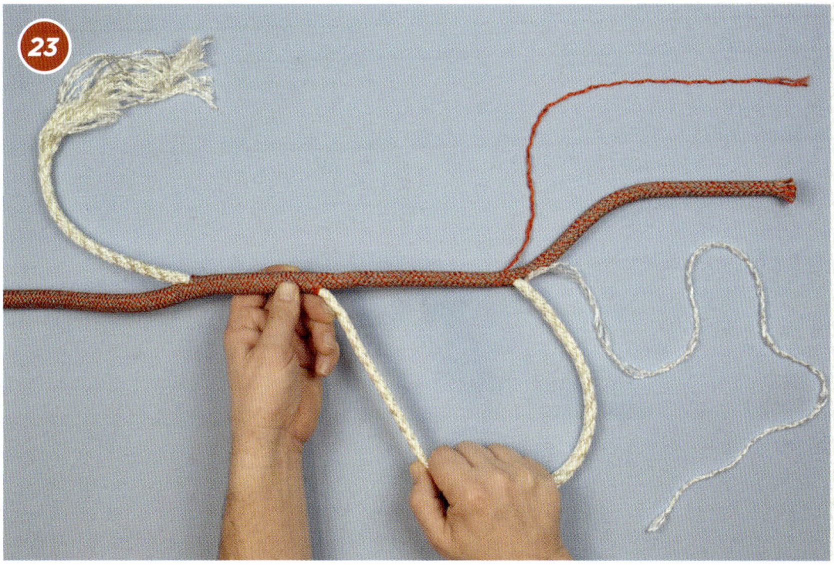

From the other side, pull out more length of the core.

Taper the cover end by pulling out single yarns one behind the other. All details about tapering can be found in the chapter "Basic Splicing Skills" on page 20.

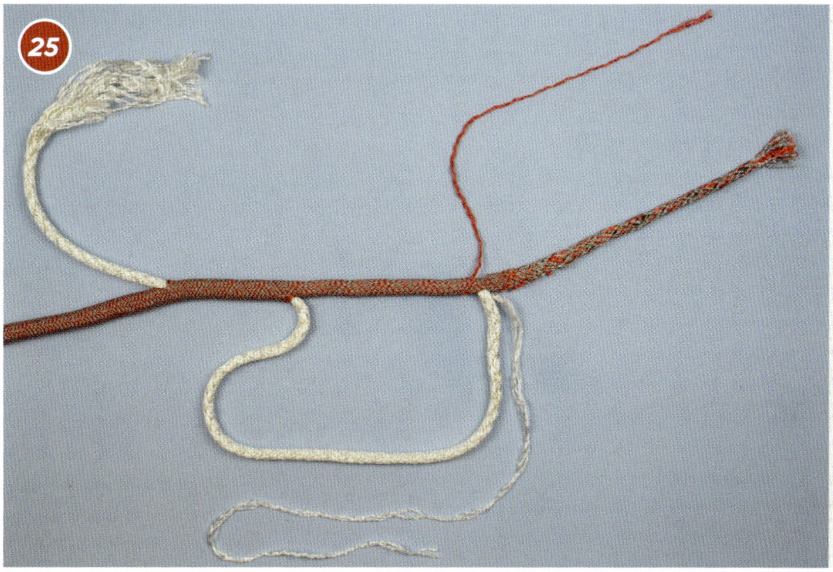

The cover end has been tapered, and its end now consists of only about half the number of yarns. Note that the braid has not been dissolved by the taper.

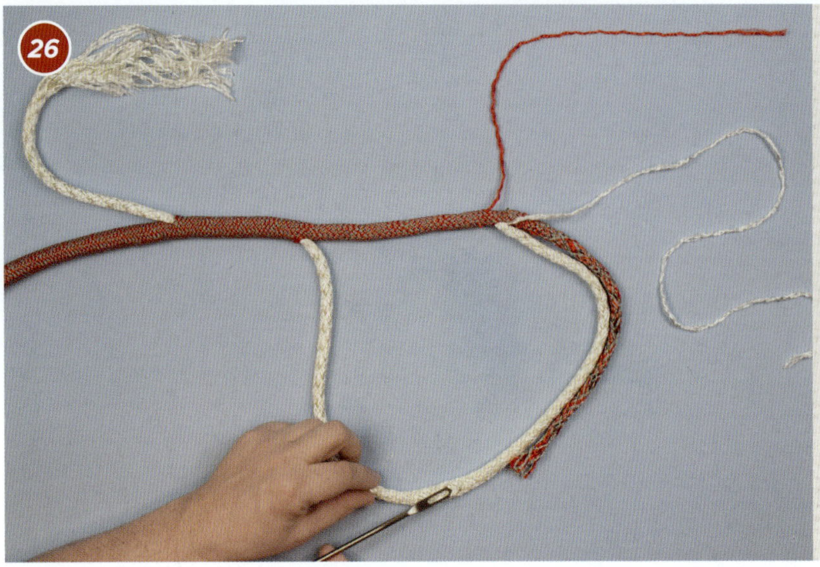

The tapered cover end is now spliced into the core. First, place the cover end parallel alongside the core. Insert the splicing needle into the core but go in a few centimeters farther away from where the cover end is, to allow for the shortening effect.

Exit with the splicing needle at the same height as the core marker yarn (white). Thread the cover end onto the splicing needle.

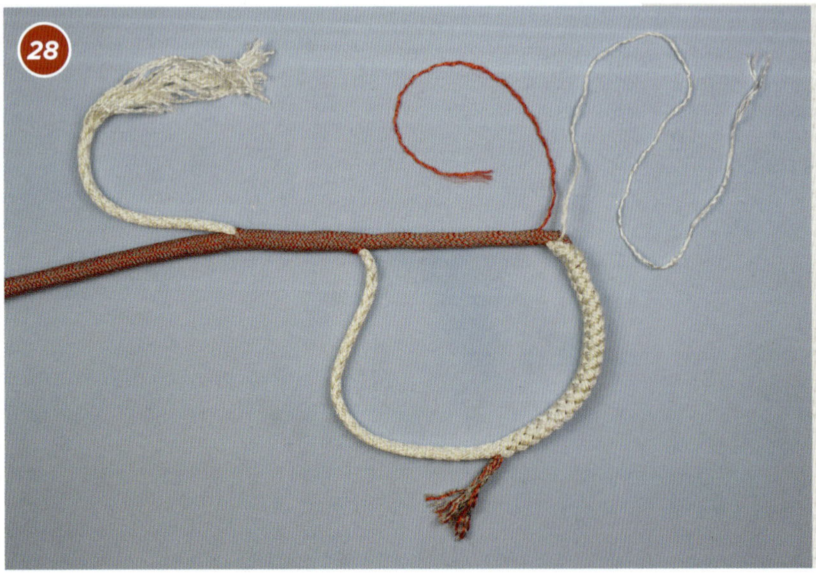

The cover end has been pulled through.

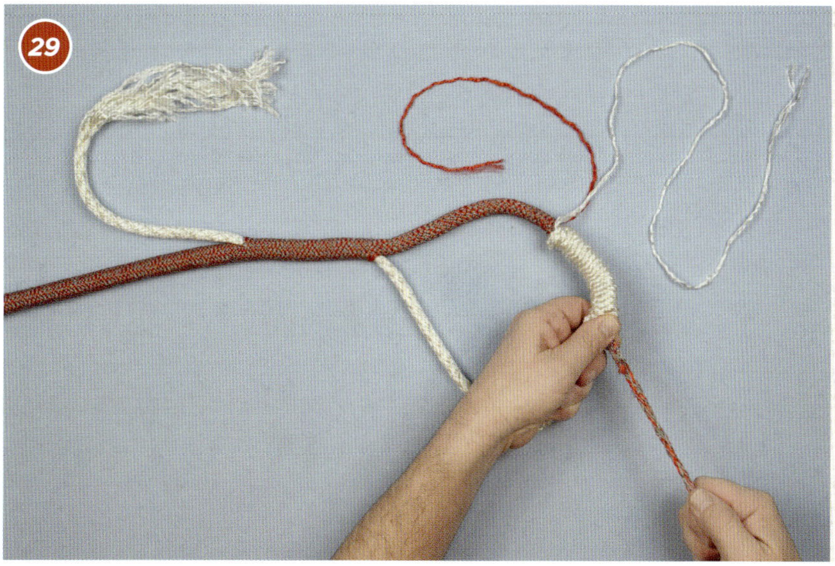

Grab the rope with one hand where the cover end emerges, and firmly pull on the cover end with your other hand.

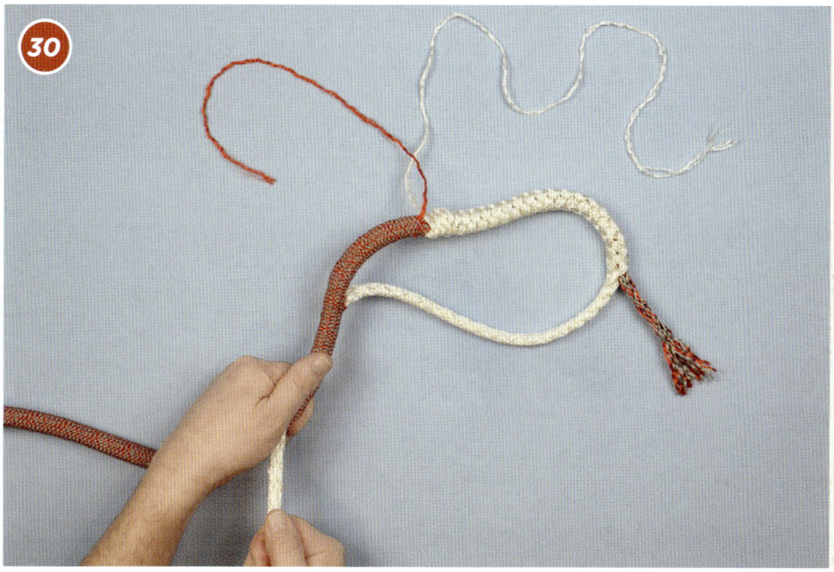

Now grab the rope with one hand where the core end emerges, and firmly pull at the core end with your other hand.

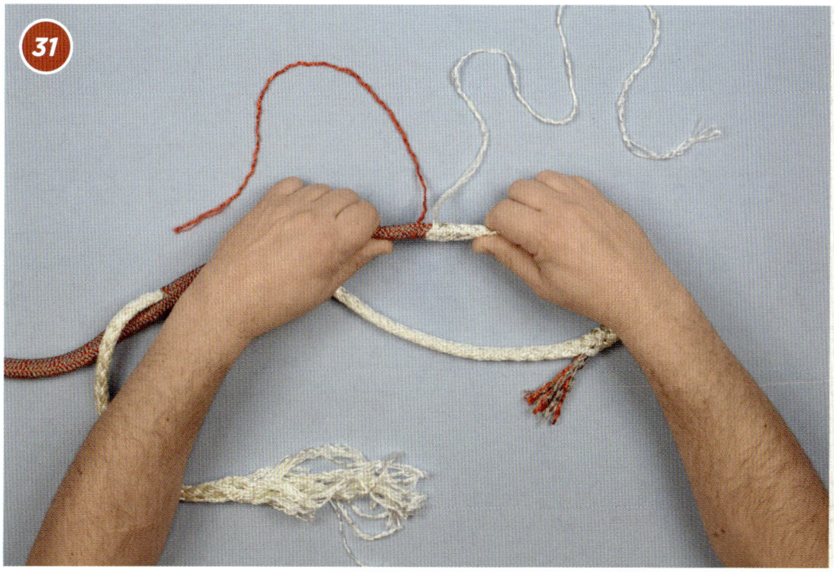

By pulling at the ends, the crossover is nicely pushed together; that is the point where the core goes into the cover and the cover goes into the core. Smooth back toward both sides, starting at the crossover, and this will give a nice transition without any undue thickness.

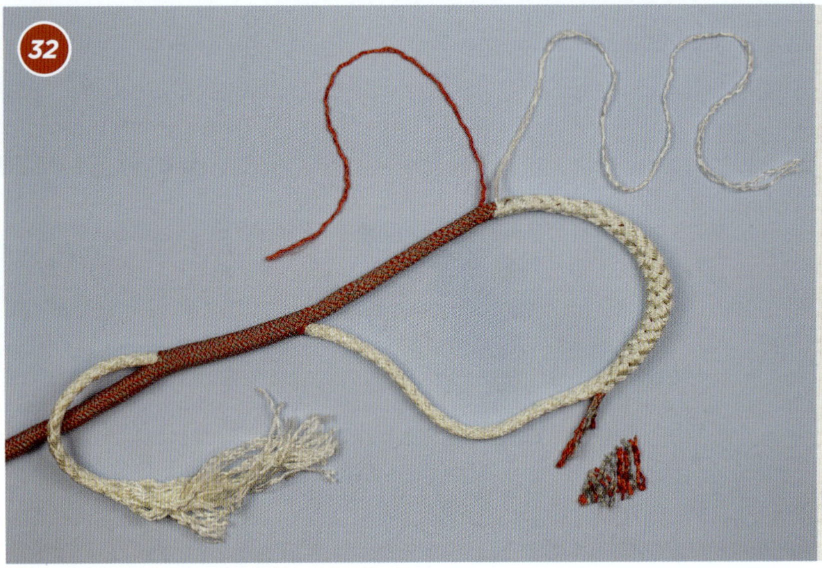

Cut the cover end at a sharp angle to bring the taper to perfection, and let the cover end disappear completely inside the core.

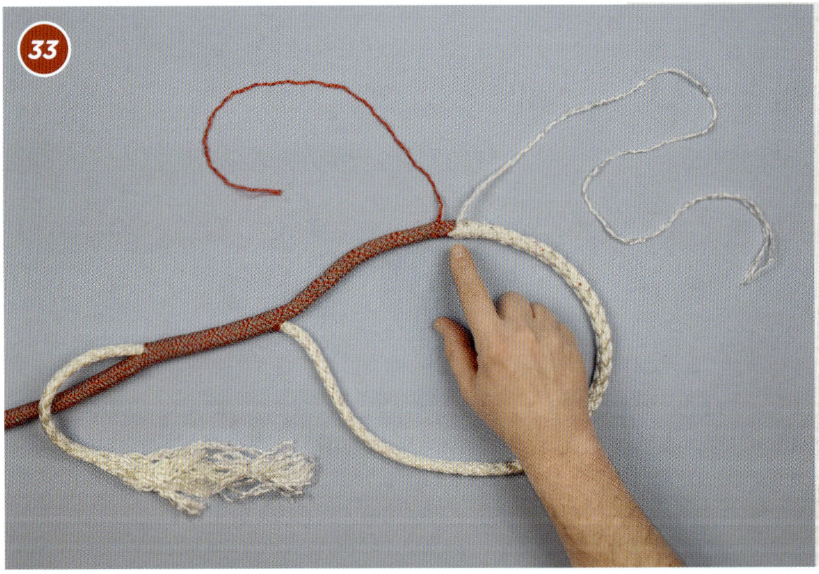

Exactly at the crossover, the core marker yarn (white) emerges.

The core marker yarn is used to secure the crossover. Thread the core marker yarn onto a sewing needle and stitch about four times across. Place the stitches close to each other and toward the side where the core (white) is on the outside. These stitches secure the crossover against any shifting or sliding, especially under the lesser loads that occur repeatedly in form of jerks or vibration.

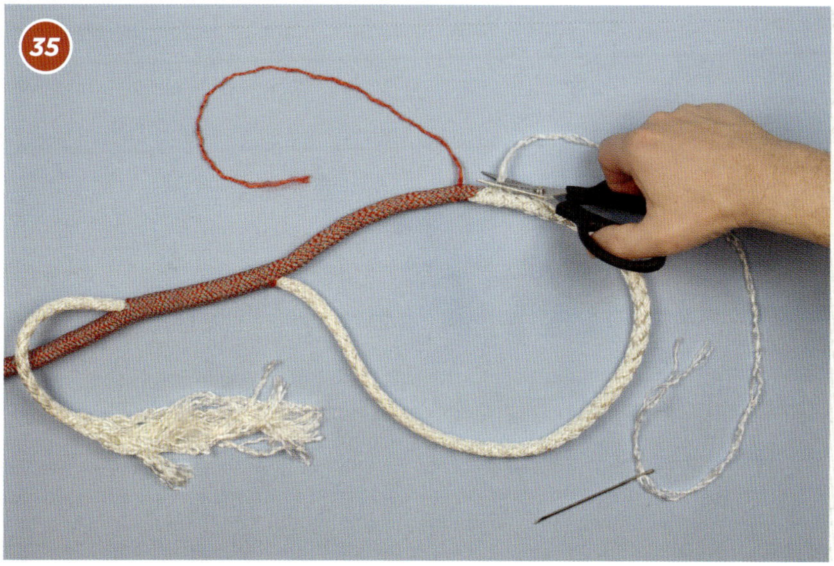

After having stitched about four times across the rope, the core marker yarn is cut off flush.

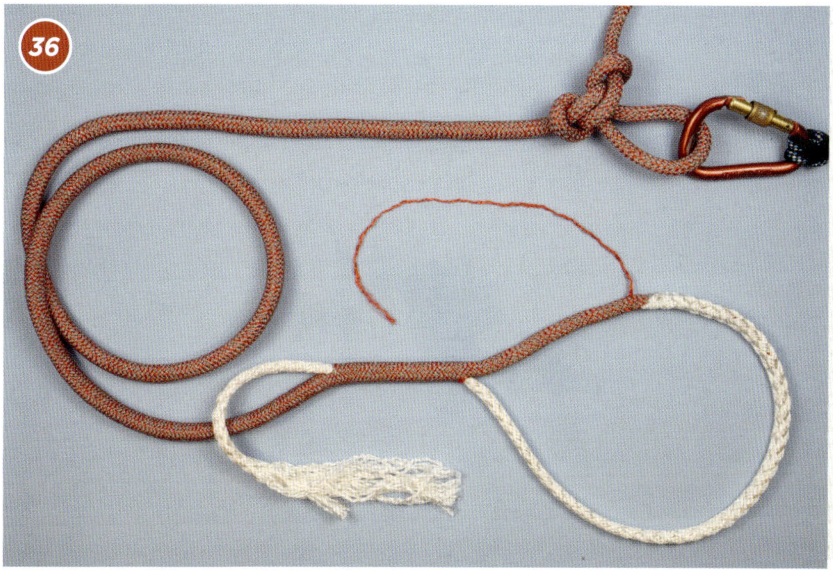

Attach the loop knot to a strong point and start milking back the bunched cover. All details about how to milk back can be found in the chapter "Basic Splicing Skills" on page 20.

Milk back until the cover marker yarn is only a few centimeters or about 2 inches from the entry spot. Before proceeding any further with milking back, the protruding end of the core needs to be taken care of.

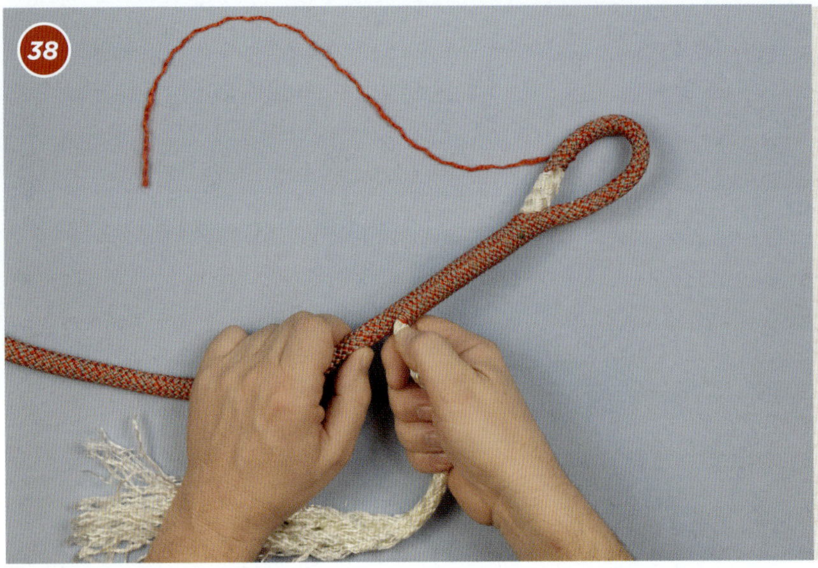

Grab the core end, firmly holding it with your thumbnail at exactly the spot where it emerges from the rope.

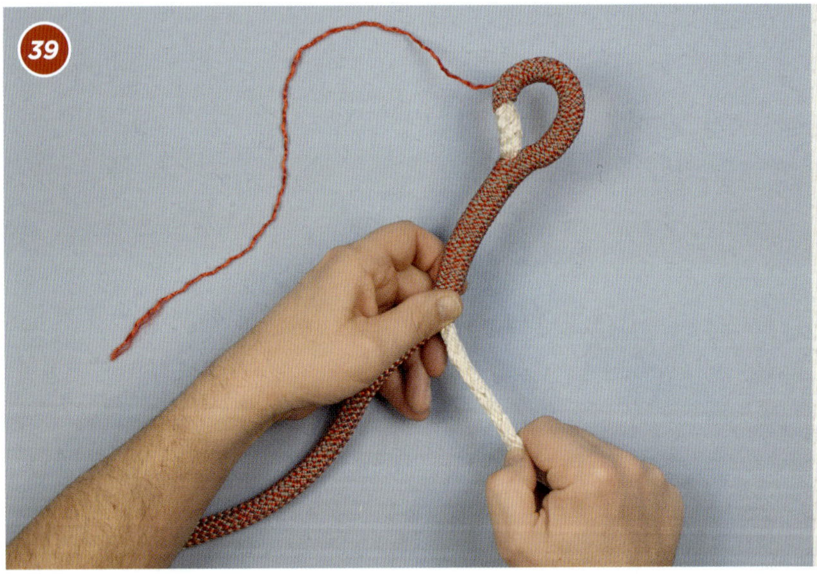

Pull out the core for a few centimeters or about 2 to 3 inches.

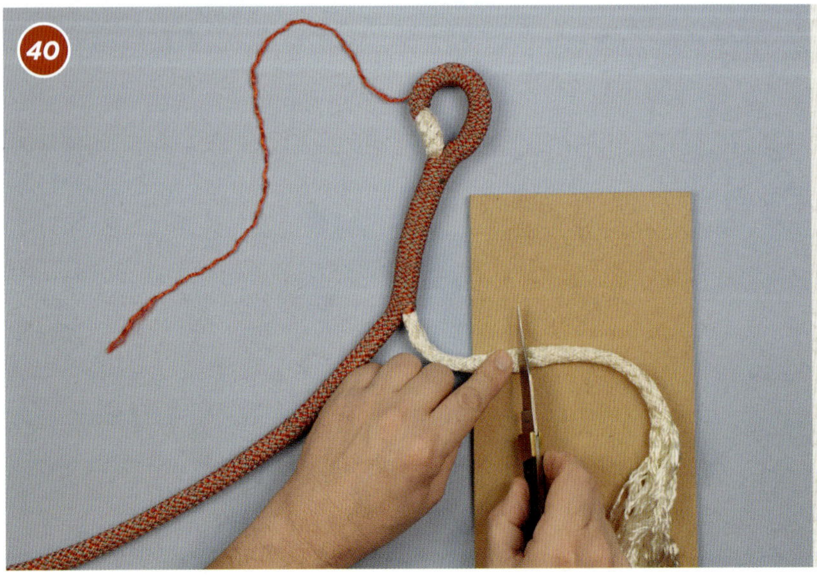

Cut the core at exactly the spot where you were holding it with your thumbnail.

Unravel the end and cut it again, this time at a sharp angle so that it tapers to a point.

Firmly smooth back the bunched cover, starting at the crossover point, going around the eye and down the rope.

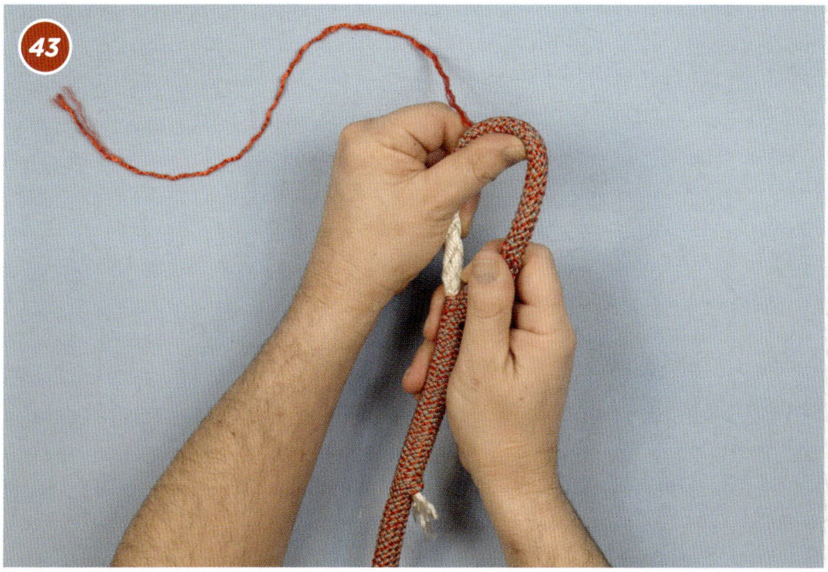

Smoothing back down the rope. The core end will disappear inside.

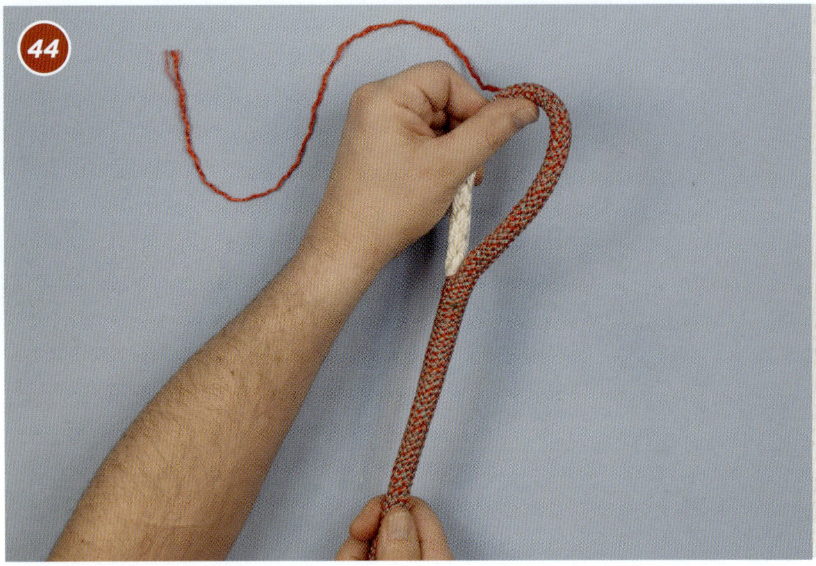

The core end has disappeared. The eye has increased in size (corresponding exactly to the distance the core end has been temporarily pulled out before). Now milking back can be resumed, and the eye will be closed completely.

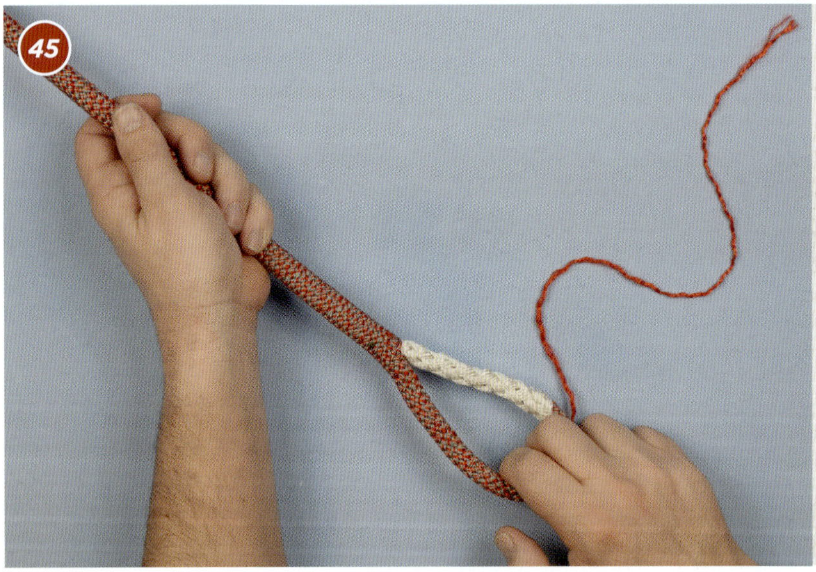

With the rope attached to a strong point by its loop knot (butterfly knot), continue to milk back the bunched cover. Milk back with one hand and—very important—simultanously hold the entire length under tension by pulling at the eye with your other hand. All details about milking back can be found in the chapter "Basic Splicing Skills" on page 20.

The cover marker yarn has reached the entry point, and the eye has exactly its predefined size.

Thread the cover marker yarn onto a sturdy sewing needle, preferably a sailmaker's needle, and make four closely spaced stitches across the throat of the splice in order to secure it when under jerking or vibrating lesser loads.

In case distances were marked with a felt pen (instead of marker yarns), a palm-and-needle whipping (stitched whipping) should be applied at the throat of the splice. Only if the spliced section is very firm can securing stitches be omitted

Done! The splice is finished. Note the following points to determine good craftmanship: 1. There are no visible or tangible steps or pronounced increments over the entire spliced section. The splice has a nice conical shape with regular increase in diameter. Careful tapering was necessary to achieve this! 2. The cover is not bunched around the eye circumference; the eye rather feels exactly like the rope in its free length. If the section around the eye does feel bunched, it would be because the entire length of the rope has not been held under sufficient tension while milking back.

2.2 EYE/THIMBLE SPLICE IN ROPES WITH PARALLEL FIBER CORE

Kernmantle ropes from polyester or other conventional fiber materials with a core of loose, parallel fibers.

The parallel-lying fibers of the core are oriented exactly in load direction. This completely averts construction stretch, which is inevitable in any kind of braid due to the winding run of strands and yarns. Ropes with parallel fiber cores have comparably little stretch even if they consist of just conventional fibers. The difference in splicing these ropes is that the cover end cannot be spliced into the core and must therefore be affixed by tape.

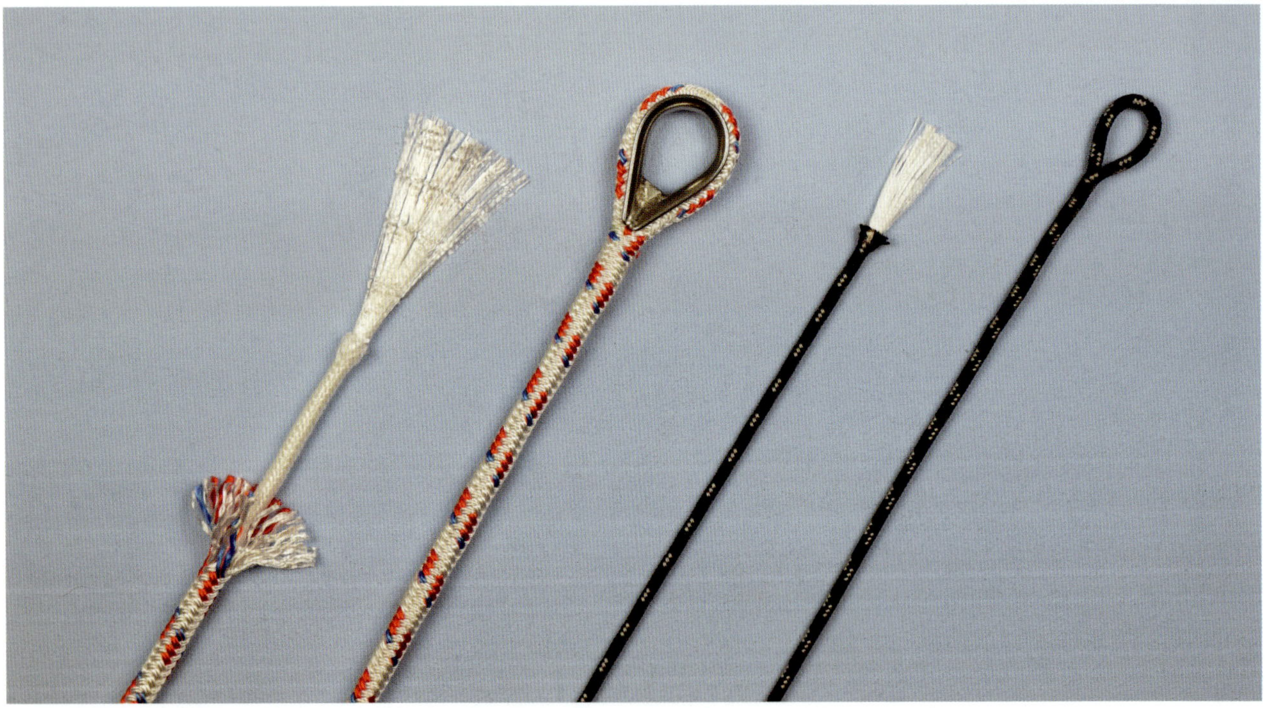

START

VIDEO 2.2

Construction:	Kernmantle rope with parallel fiber core
Material:	core: polyester, cover: polyester
Difficulty:	medium
Splicing tools:	splicing needle or similar pull-through tools

Ropes with parallel fiber cores often have an intersheath bundling together with the parallel-lying fibers of the core. This intersheath is removed over the entire length of the splice. If no such intersheath is present, steps 3 and 4 can be skipped.

Tie a loop knot, preferably a butterfly knot, about 3 meters / 10 ft. from the end of the rope. Cut the end of the rope with a knife; it must not be fused. Measure the splicing length of the cover. This length is 14 times the rope's diameter. Here, 14 x 10 mm diameter = 140 mm or 14 cm. At the measured distance, pull out one single yarn from the cover braid to mark this spot. This is the cover marker yarn.

From the cover marker yarn, measure the desired circumference of the eye splice. If a thimble is to be spliced into the eye, lay the rope around the thimble to pick off the eye circumference needed. The circumference starts at the cover marker yarn. Mark the end of the eye circumference with a felt pen.

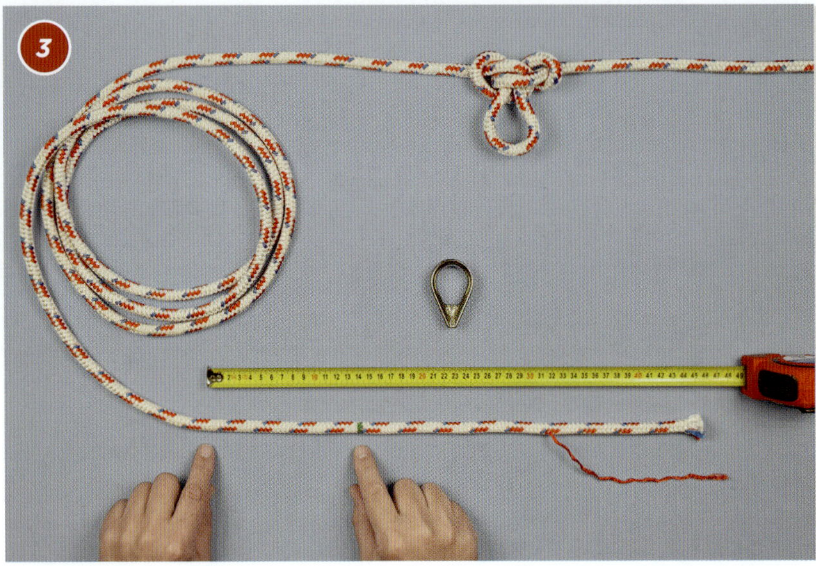

From the spot you marked with the felt pen, measure again 14 times the rope's diameter; here, 14 cm. From there, the total distance to the rope's end now amounts to 45 cm. This is the length over which the intersheath has to be removed. If there is no intersheath present in the rope, continue with step 5.

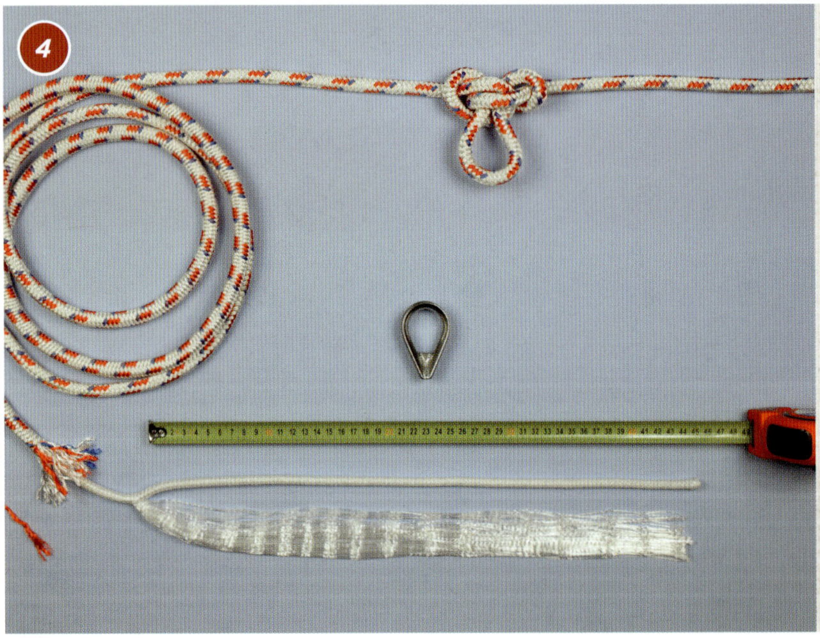

Pull out the core, including the intersheath, from the end of the rope by pushing back the cover. Measure the total length (here, 45 cm) from the end of the core. At this spot, the intersheath has to be cut. In order not to compromise the core or accidentally cut single fibers of the core, it is often the better choice to extract the core at the measured spot from the intersheath and then cut the hollow intersheath right next to the extracted core.

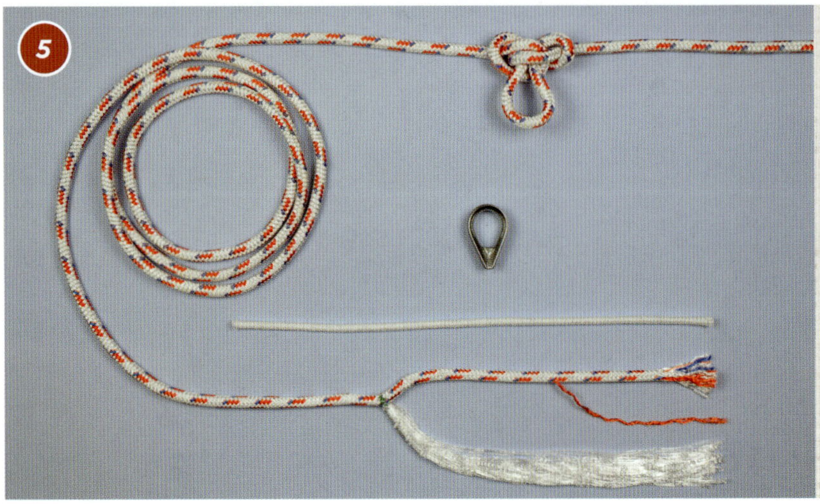

Milk back the cover as far as possible, starting at the butterfly knot. Then carefully extract the core at the spot marked with the felt pen.

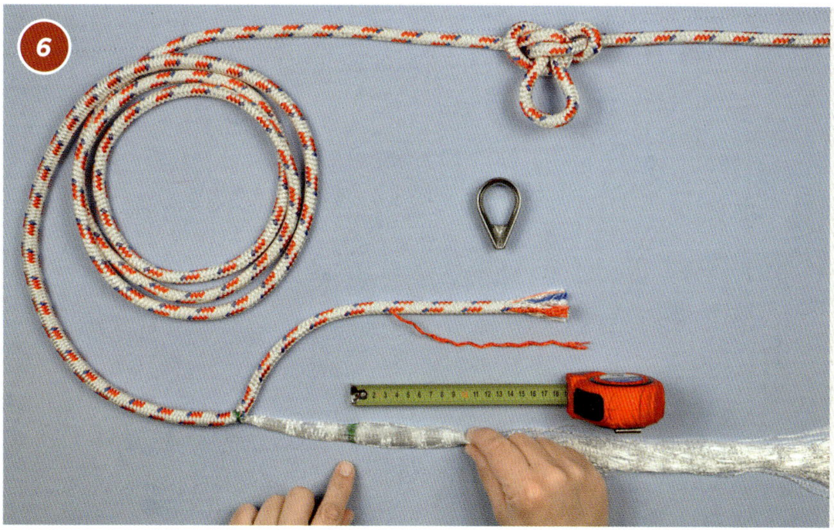

Grab the core where it emerges from the cover, and pull it out farther. From the spot where you are holding the core, measure the core shortening, which is 8–10 times the rope's diameter. Mark this spot on the core with a felt pen.

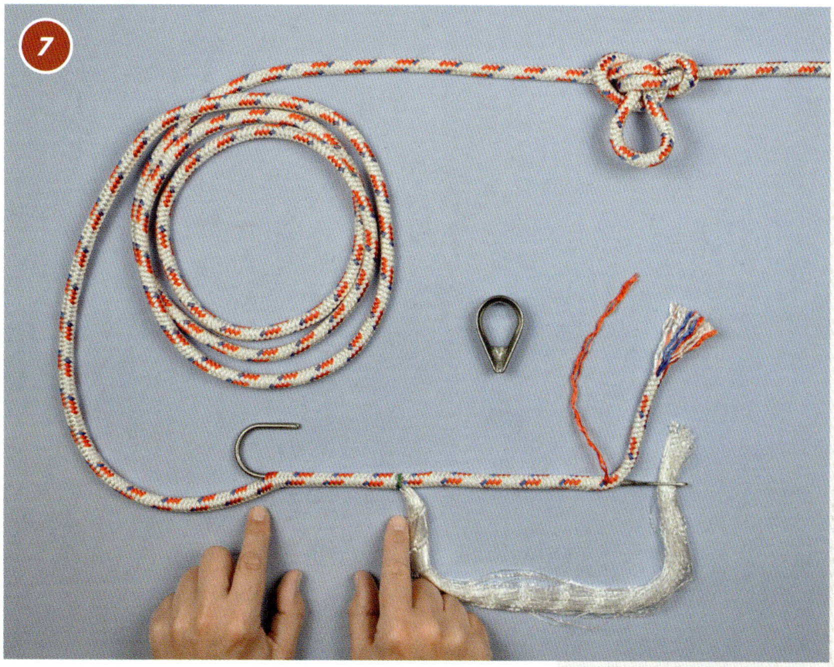

Again milk back the cover as far as possible, starting from the butterfly knot. The marked spot on the core will be pulled back inside. Measure 14 times the rope's diameter from the spot where the core emerges (pointed out in the picture). This gives the spot where to insert the splicing needle. This is also the spot where the intersheath ends inside the rope. Push the splicing needle through and bring it out one rope diameter behind the cover marker yarn. Thread the core onto the splicing needle.

Pull the core through, and at the same time pull out more of the core from the side of the standing end. The spot marked with the felt pen will reappear again.

Continue pulling the core through until the spot marked with the felt pen reaches the entry point into the cover. Also pull out more of the core from the other side, the side of the standing end. Taper the cover end and cut the very end at a sharp angle. All details about tapering can be found in the chapter "Basic Splicing Skills" on page 20.

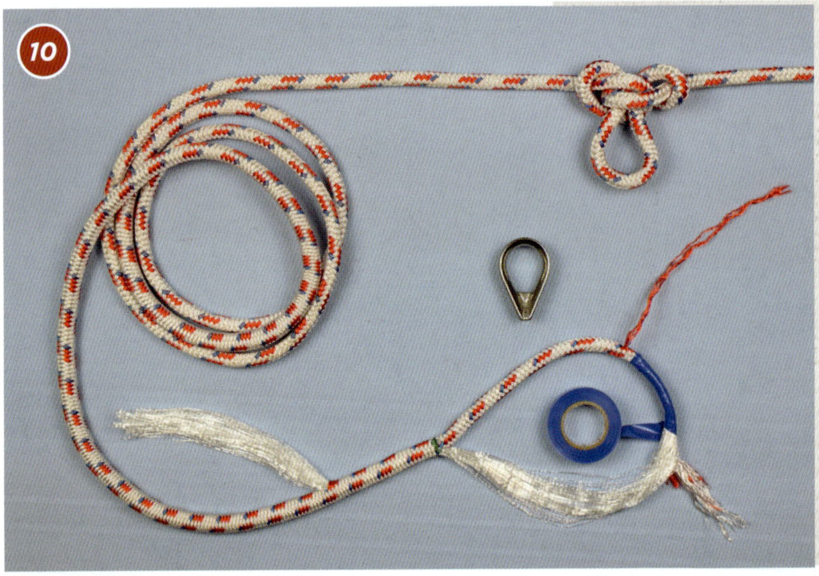

The single yarns pulled out for the taper are cut off flush, and the cover end is then affixed to the core with a thin, tightly wrapped layer of tape (e.g., PVC electrical tape).

Now milking back can be started. Attach the line with the butterfly knot to a strong point. All details about milking back, including tips and tricks, can be found in the chapter "Basic Splicing Skills" on page 20.

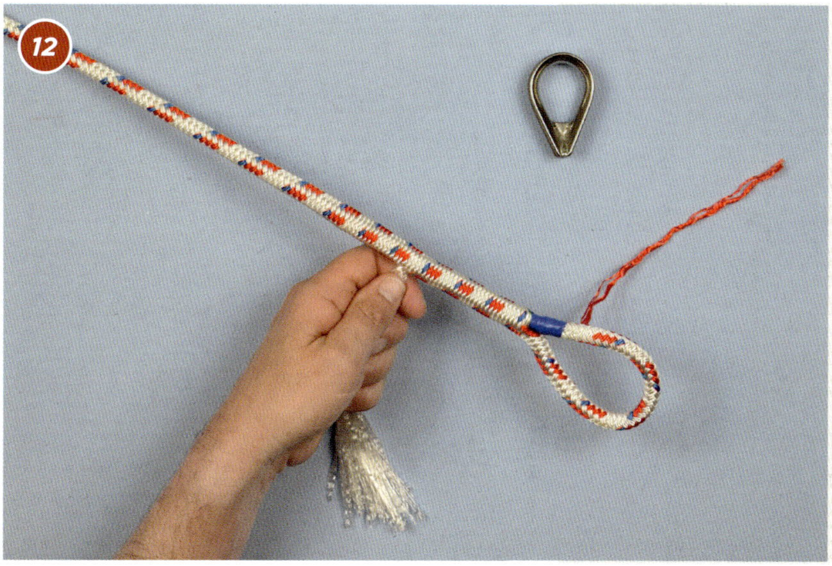

The wrapped section is pulled inside and the eye closes. Before closing the eye splice completely, take care of the core end, which still protrudes from the rope. Firmly grab the core exactly where it emerges, and …

… pull the core out a little farther. Cut the core at exactly the spot where you were holding it.

Cut the very end of the core again, but this time at a sharp angle to give it a little taper.

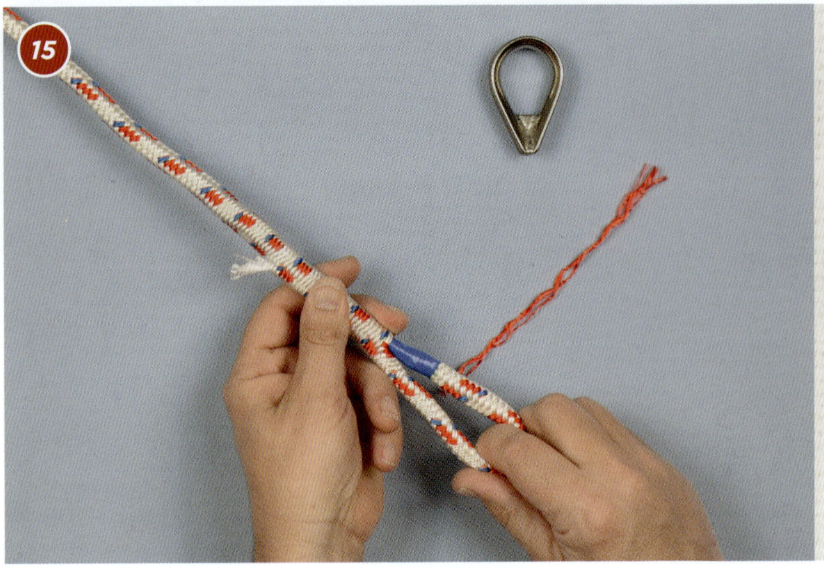

Smooth back firmly from the side of the eye, and the core end disappears inside.

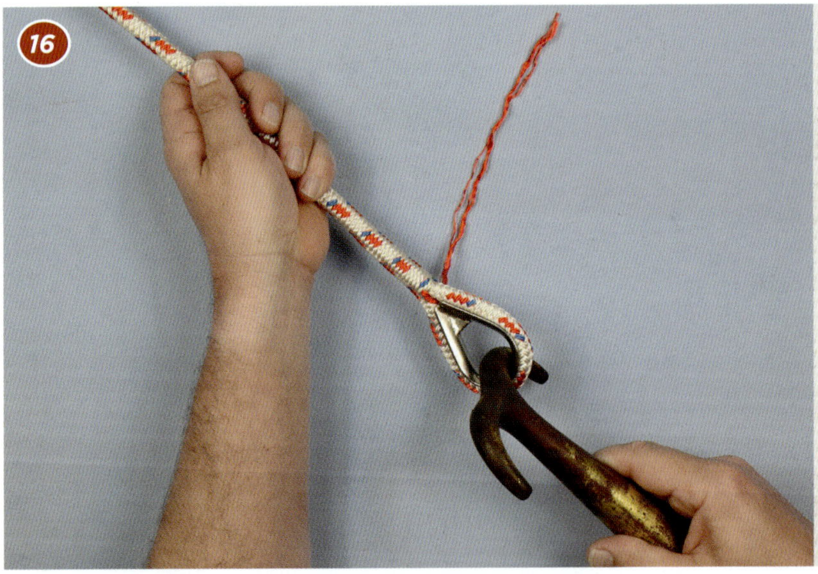

Now the eye splice can be closed completely. The rope needs to be attached to a strong point at the butterfly knot. Milk back with one hand and—very important—simultanously pull at the eye with your other hand, holding the entire length under tension at all times. If a thimble is to be spliced into the eye, insert the thimble at the last moment, before the eye closes too far.

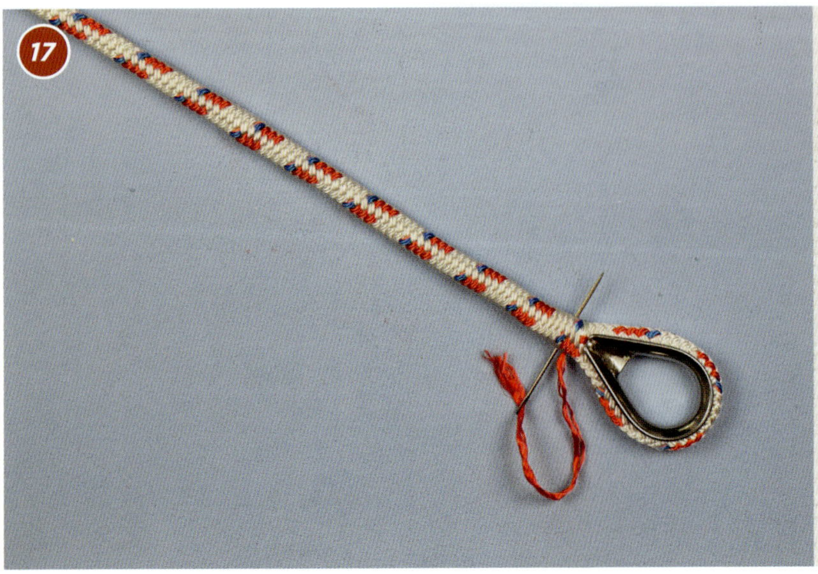

Finally, secure the splice with the cover marker yarn by making about four closely spaced stitches across the throat of the splice.

2.3 END-TO-END SPLICE IN CONVENTIONAL FIBER ROPES

Double-braided ropes with core and cover from polyester (PES) or polyamide (PA) or polypropylene (PP). The splicing technique for these conventional fiber ropes can in principle also be used for ropes with a core of high-modulus fibers such as Dyneema or Spectra if maximum breaking strength is not of primary concern; compare chapter 2.5.

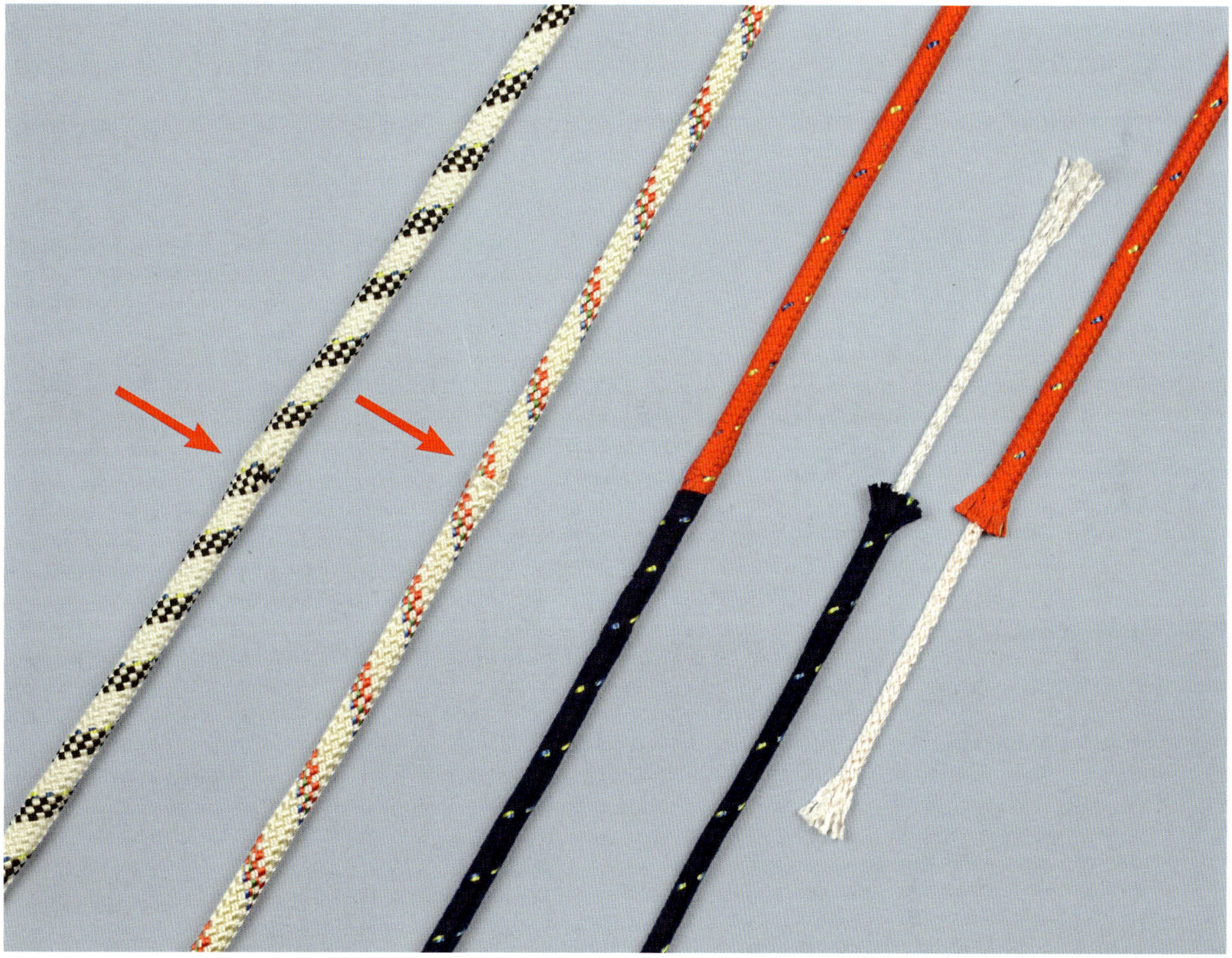

End-to-end splices often have to be made directly on board because furling lines need to be roved to furling gear before making the splice. A well-made end-to-end splice is hardly visible, and it requires a close look to discern the spot where the ends of the line have been joined. In principle the same technique can also be used to join the ends of two different ropes.

1. Method: End-to-end splice with maximum breaking strength

This method results in a small increase in diameter at the spliced section.

START VIDEO 2.3

Construction:	double braided, kernmantle rope
Material:	core: polyester, cover: polyester
Difficulty:	medium
Splicing tools:	splicing needle or fid

Tie a loop knot, preferably a butterfly knot, at about 3 meters / 10 ft. from each end of the rope. The ends need to be cut with a knife and must not be fused.

Measure a distance of 24 times the rope's diameter from the end of the line; here, 24 x 10 mm diameter = 240 mm or 24 cm. Mark this spot with a felt pen. From now on, all steps performed on one end of the rope are to be replicated in exactly the same way on the other end of the rope.

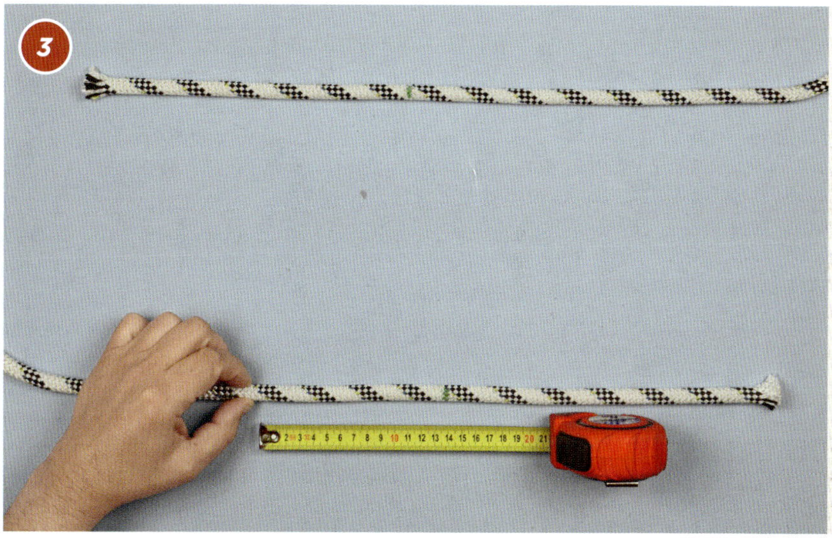

Measure from the marked spot 14 times the rope's diameter and extract the core at this spot. How the core is extracted can be found in the chapter "Basic Splicing Skills" on page 20.

Grab the core at exactly the spot where it emerges, and pull it out farther.

From the spot where you are holding the core, measure the core shortening, which is 15–18 times the rope's diameter; here, 15 x 10 mm diameter = 150 mm or 15 cm. At this spot, pull out one single yarn. This will be the core marker yarn.

***TIP:** The exact length of the core shortening varies slightly between different rope types and manufacturers and depends on the type of braid and the ratio of volumes between core and cover. If in doubt, use 18 times the rope's diameter to make sure you can completely close the end-to-end splice when finishing the work.*

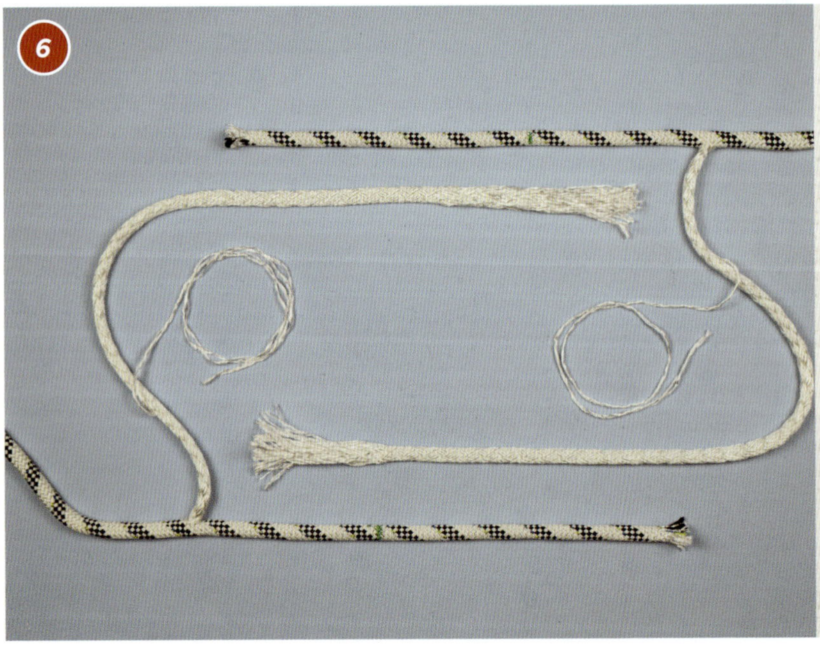

Both core marker yarns have been pulled out of the cores. The cores are placed parallel to the opposing hollow cover ends. Each core will now be spliced into the section of the cover that is 14 times the rope's diameter; in other words, from the spot marked with the felt pen to the spot where the other end of the core has been extracted.

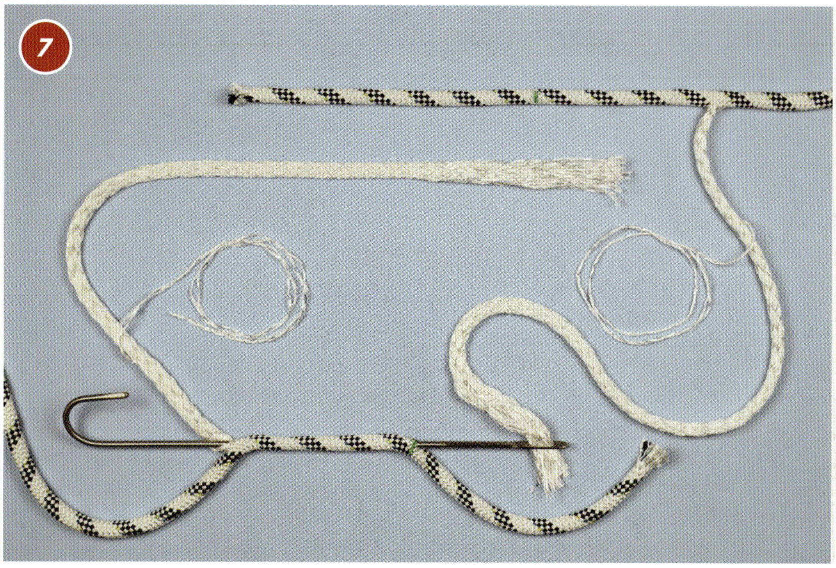

7

Insert the splicing needle at exactly the opening where the core emerges. Push the splicing needle through the hollow cover end and bring it out at the spot marked with the felt pen. Thread the core onto the splicing needle and pull it through.

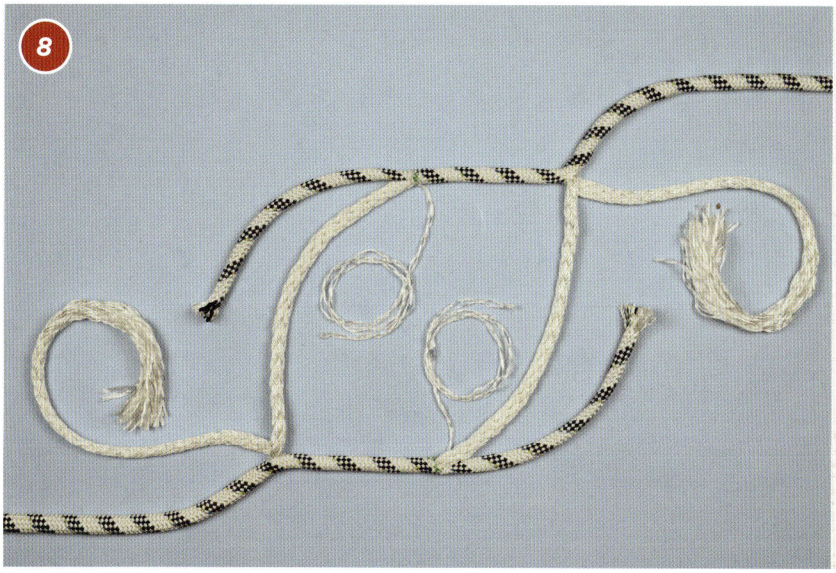

8

Both cores have been spliced into the cover ends. Pull the cores through until the core marker yarns are exactly at the entry points, as in the picture.

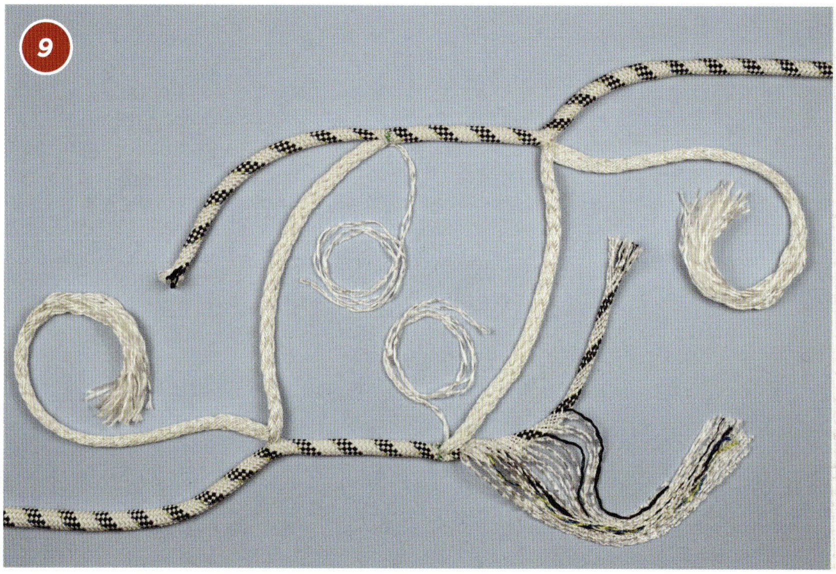

9

Taper the cover ends from where the core enters by pulling out single yarns one behind the other. All details about tapering can be found in the chapter "Basic Splicing Skills" on page 20.

Insert the splicing needle into the core at a sufficient distance to pull the cover end into the core. Bring the splicing needle out at the same height as the core marker yarn. Thread the tapered cover end onto the splicing needle and pull it inside the core.

Both cover ends have been pulled into the cores. Make sure the crossover spots where the marker yarns emerge are closely pushed together. Before letting the cover ends disappear inside the cores, cut the very ends at a sharp angle.

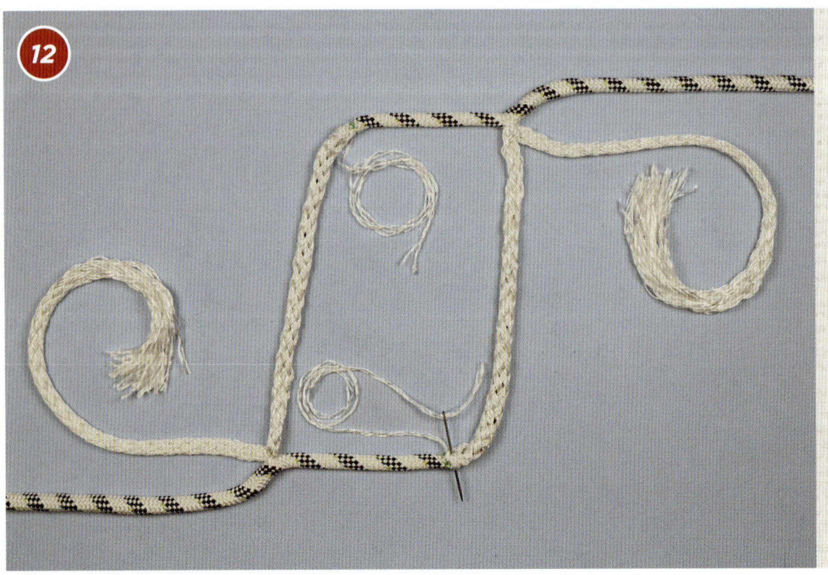

Smooth back over the cores to let the cover ends completely disappear inside. Secure the crossover spots with the core marker yarns. Make about five stitches across the line and then cut the core marker yarns off flush.

The end-to-end splice is ready to be closed. Alternately attach the rope with one of the two butterfly knots to a strong point and start milking back. All details about milking back can be found in the chapter "Basic Splicing Skills" on page 20.

The splice is closed step by step. Attach the rope with one of the butterfly knots and milk back. Then attach the rope with the other butterfly knot and milk back from the other side. Keep alternating milking back from one side and then from the other side. Here the splice is about to "swallow" the cross-over spots.

Now only a few centimeters are still missing.

Before the splice closes all the way, cut the protruding core ends off flush. Milking back over the splice from both sides will make the ends disappear inside.

The finished end-to-end splice. At a glance, it is not easy to even make out the spot where the ends are joined.

2. Method: End-to-end splice without increase in rope diameter

The same end-to-end splice can also be made without any increase in diameter by augmenting the taper of the core and cover ends. That means more yarns will be pulled out, and the taper will extend over a greater length. As a drawback, the breaking strength will be slightly diminished.

This second method differs only in steps 9–11 from the first method shown above. Proceed to step 8 as shown above. Only in step 5, measure a lesser distance for the core shortening of only eight times the rope's diameter.

START VIDEO 2.3

Construction:	double braided, kernmantle rope
Material:	core: polyester, cover: polyester
Difficulty:	a little more challenging
Splicing tools:	splicing needle or fid

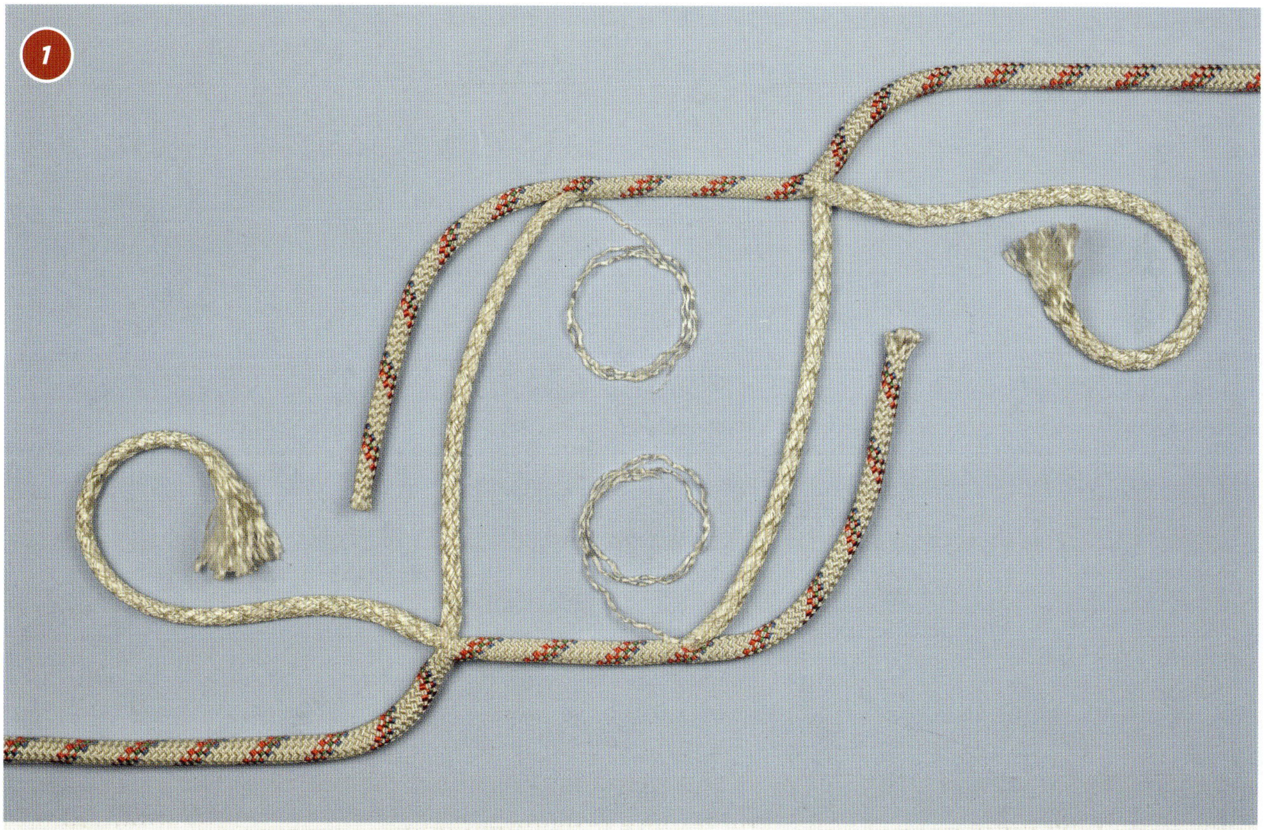

Up to this step, everything is the same as with the first method (end-to-end splice with maximum breaking strength); compare step 8 from above, only that this time the core shortening was measured with eight times the rope's diameter. The core ends have been pulled through to the point that the core marker yarns are at the entry spots into the cover. From now on, all steps performed on one end of the rope are to be replicated in exactly the same way on the other end of the rope.

The cover end is now tapered over its full length; that means the taper starts at the spot where the core was extracted. Work very carefully when making the taper, and pull out about half of the yarns of the cover braid. Pull out the yarns very close to each other to confine the taper to a short length. The yarns should be pulled out on the same side on which the core emerges. This side will be on the inside once the splice is closed. All details about tapering can be found in the chapter "Basic Splicing Skills" on page 20.

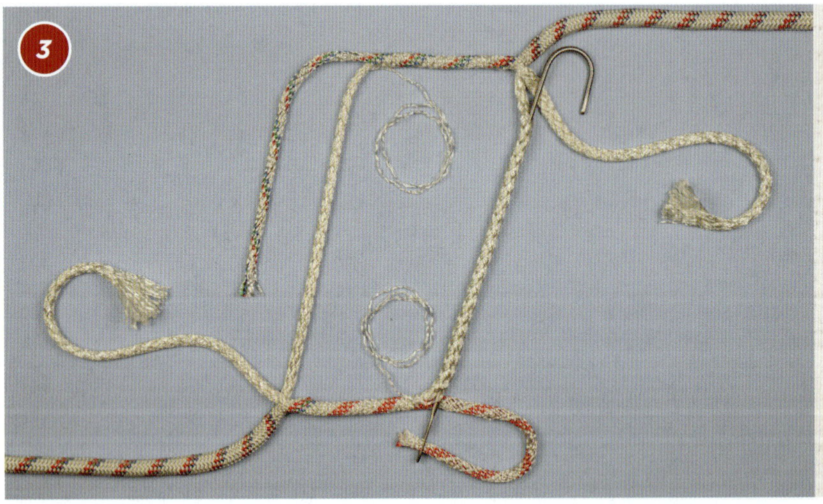

The tapered cover end is spliced into the core.

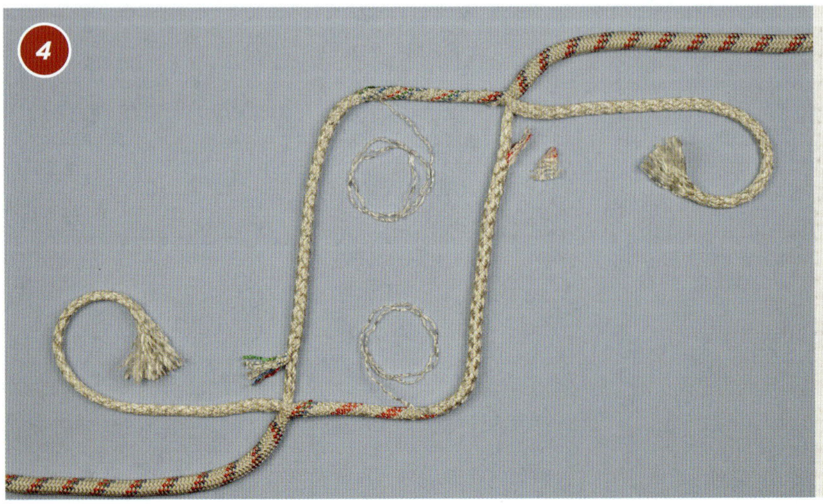

Both cover ends are inside the cores. Make sure the crossover points where the core marker yarns emerge are tightly pushed together. Cut the very end of the covers at a sharp angle before letting them disappear inside the cores.

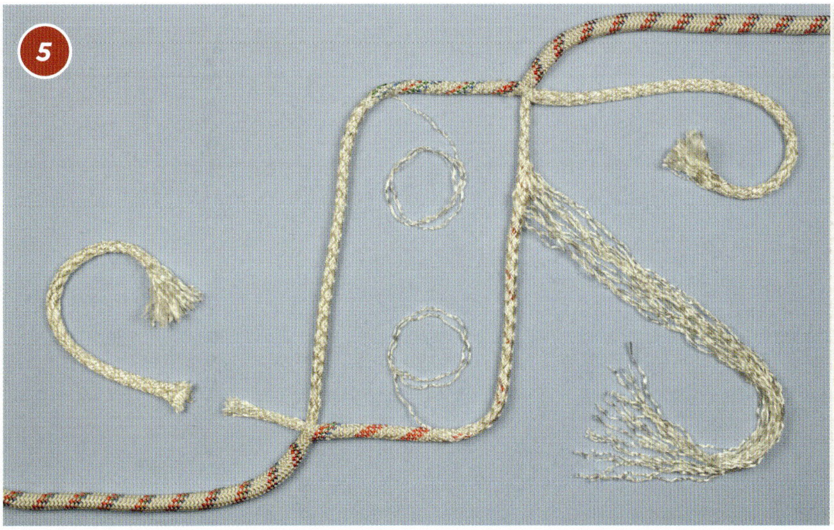

Now the cores need to be tapered also. First, cut most of the protruding end off (*left*) so that the yarns won't have to be pulled out over an unneccessarily long distance. Pull out about half the number of yarns that the core consists of. Start the taper where the cover terminates inside the core.

Both cores have been tapered, and the yarns that were pulled out have been cut off flush. The crossover points are secured by stitching through with the marker yarns. Make about five stitches across the rope and then cut the marker yarns off also.

Now the end-to-end splice is ready to be closed. Attach the rope alternately with one of the butterfly knots to a strong point and start milking back. All details about milking back can be found in the chapter "Basic Splicing Skills" on page 20.

The final steps while milking back are again identical to the first method (end-to-end splice with maximum breaking strength). Refer to steps 14–16 of the first method to view these final steps in more detail.

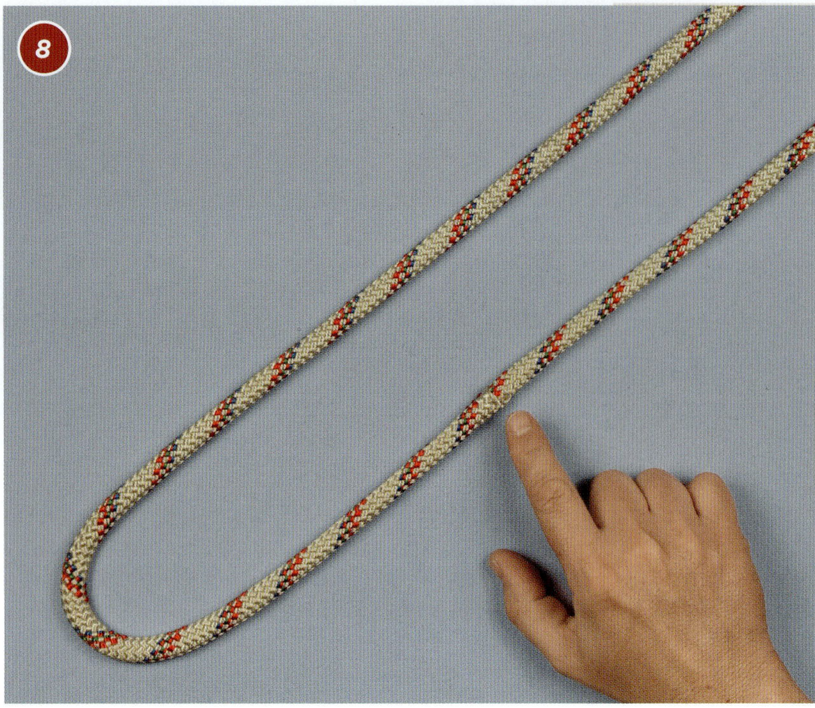

The finished end-to-end splice. The rope is of uniform diameter througout its entire length; the splice is hardly visible.

3. Method: End-to-end splice, simplified method (possible without increase in diameter)

Endless furling lines of smaller diameter do not have to take very high loads. These lines are controlled by hand to furl gennakers, code 0, or other foresails. With the following simplified end-to-end splice, only the cover ends are spliced inside each other, while the core ends are cut off. Breaking strength is diminished but will easily remain sufficient for any line controlled by hand. This simplified method can also be used for small-diameter, low-stretch furling lines with a core of Dyneema or Spectra.

START VIDEO 2.3.1

Construction:	double braided, kernmantle rope
Material:	core: polyester or HMPE such as Dyneema or Spectra, cover: polyester
Difficulty:	easy
Splicing tools:	splicing needle or similar pull-through tools, fids are of limited use only.

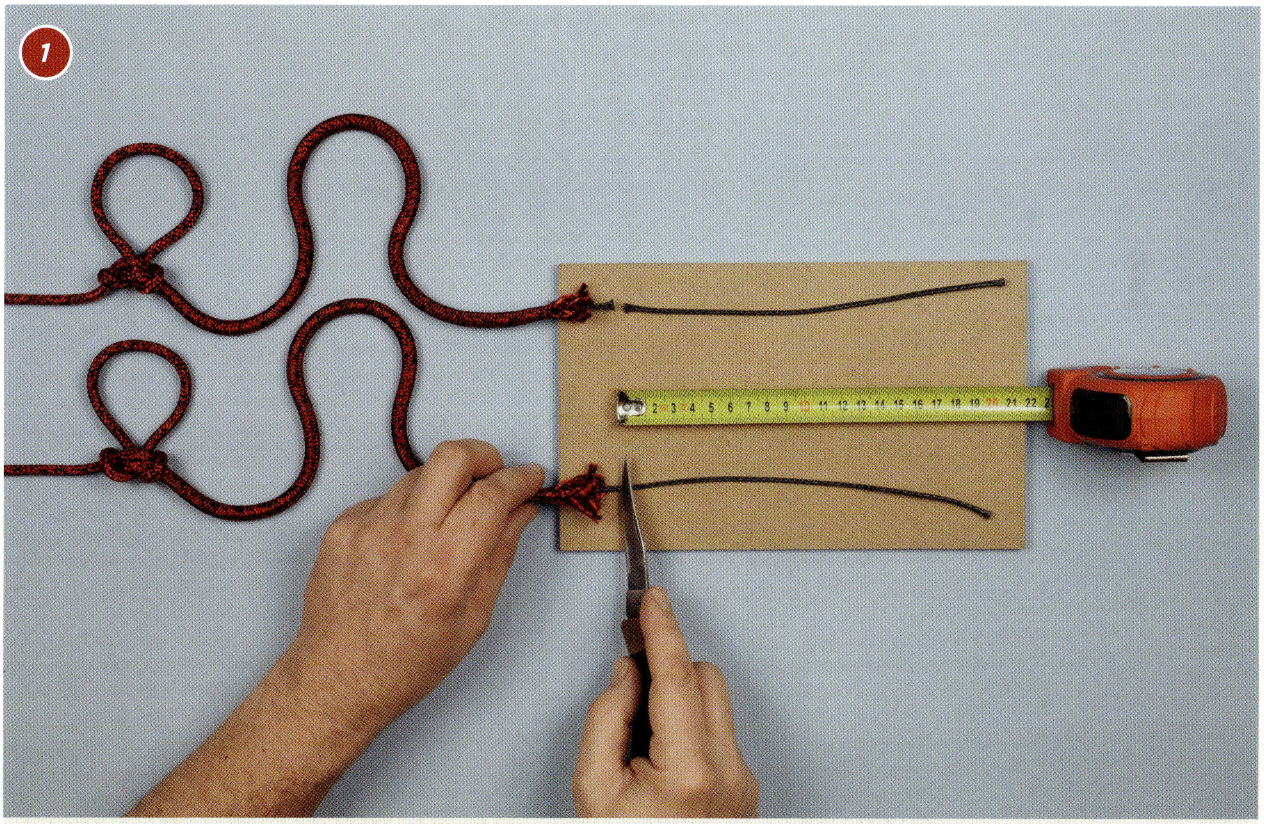

Tie a loop knot about 1 meter / 3 ft. from each end. Pull out the core from the ends of the rope by pushing back the cover. On the core ends, measure a length of 40 times the rope's diameter (here, 40 x 5 mm diameter = 200 mm or 20 cm) and cut both core ends off at this spot.

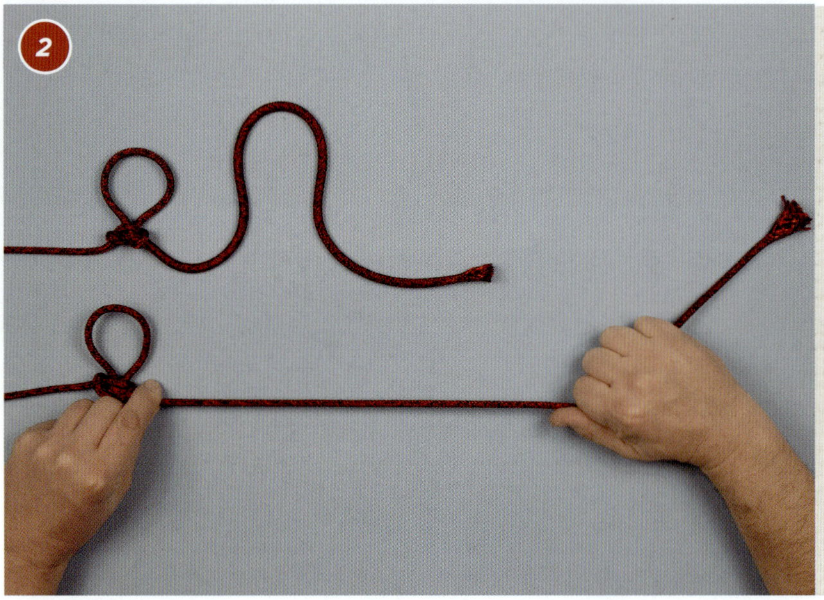

Smooth the cover back firmly, starting at the loop knot. The rope now has a section of hollow cover braid at its ends.

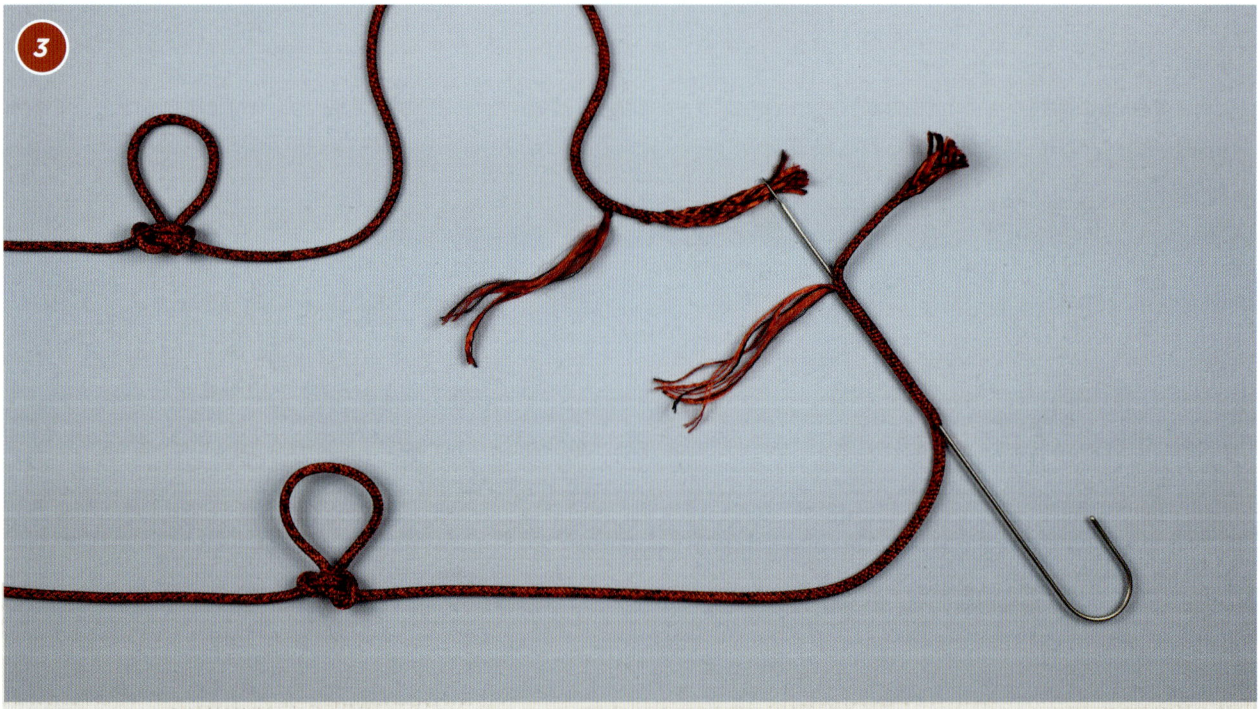

Find the middle of this length of hollow cover braid and pull out one or several single yarns, to mark this spot and to produce a small taper by removing more than one single yarn in case the cover braid is very firm.

Insert the splicing needle at exactly the spot where the core terminates inside the cover, and exit with the needle at the same height as the marker yarns. Thread the opposing cover end onto the splicing needle and pull it through until its marker yarn reaches the entry spot.

TIP: If the furling line must not have any increase in diameter at the splice, the cover ends usually need to be slightly tapered. In this case, more than one single yarn needs to be pulled out at the midpoint of each hollow cover end, as is also shown in the picture.

4

One cover end has been pulled inside the other. Now insert the splicing needle at the other end at exactly the spot where the core terminates on the inside. Push the splicing needle along the hollow cover end and exit at the same height as the marker yarns. Thread the other end onto the splicing needle and pull it inside.

5

Firmly pull on both ends so that the crossover spot is nicely pushed together.

6

Now firmly smooth back from the crossover spot toward both sides.

Cut the last protruding bits off and firmly smooth back again until the ends have completely disappeared inside.

Secure the splice with the marker yarns by making several stitches from the crossover spot toward both sides.

No end in sight: the finished furling line

TIP:* *For ropes with a diameter of more than 8 or 10 mm, it is best to have the core ends overlap the cover ends on the inside when chosing an end-to-end splice in a simplified version; see chapter 2.5, second method.

2.4 EYE/THIMBLE SPLICE IN HIGH-MODULUS FIBER ROPES

Double-braided ropes (kernmantle ropes) with a core from high-modulus fiber such as Dyneema, Spectra, Vectran, PBO, and other high-tech fibers.

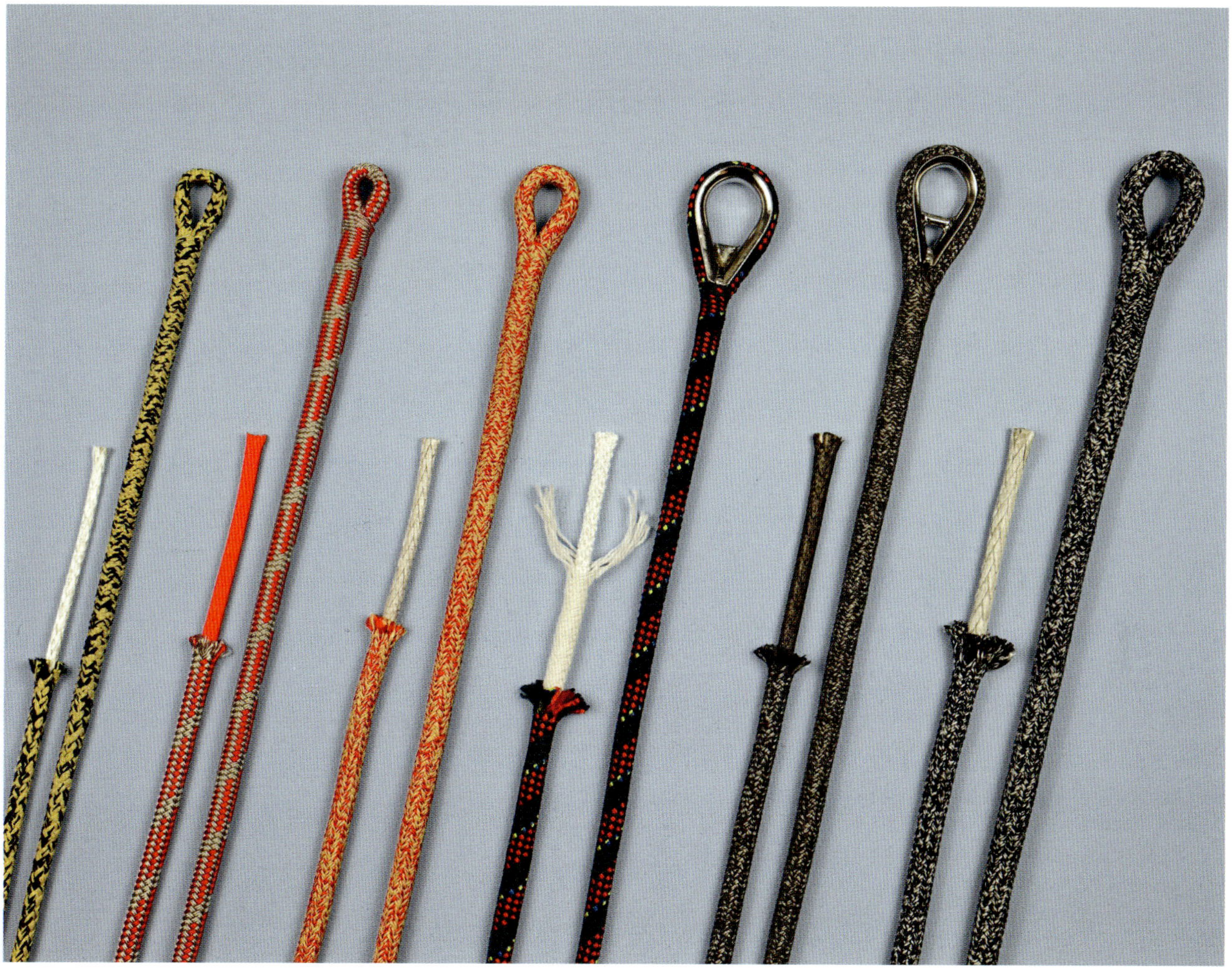

When splicing a rope with a high-modulus fiber core such as Dyneema or Spectra, the core needs to be spliced to the core. In this type of rope, only the core carries the load; the cover is there only for protection and does not carry the load. This is the reason why an eye splice for a double-braided rope with a high-modulus fiber core is fundamentally different than for a double-braided rope with a core of conventional fiber such as polyester.

START

VIDEO 2.4

Construction: double braided (kernmantle rope)
Material: core: high-modulus fiber (e.g., Dyneema, Spectra), cover: polyester or polyester mixed with high-modulus fibers
Difficulty: a little more challenging
Splicing tools: splicing needle or fid

Some of these ropes additionally have an intersheath, often made of polyester staple yarn or other materials creating high friction. The intersheath's job is to prevent the core from slipping inside the cover, especially when the rope is under high load and held by stoppers or rope jammers. In ropes without an intersheath, the necessary friction between core and cover is created by a special coating of the core. In the area of the splice, however, the intersheath is removed. As a side effect, removing the intersheath also makes for a little more space inside the cover, so that at the end of the splicing work it will also be easier to milk the thickened spliced section back into the cover.

Ropes with an intersheath: Tie a loop knot, preferably a butterfly knot, about 3–4 meters / 10–15 ft. from the end of the rope. Pull the core, including the intersheath, from the end of the rope by pushing back the cover.

Ropes without an intersheath: Tie a loop knot, preferably a butterfly knot, about 3–4 meters / 10–15 ft. from the end of the rope. Skip the next steps and continue with step 6.

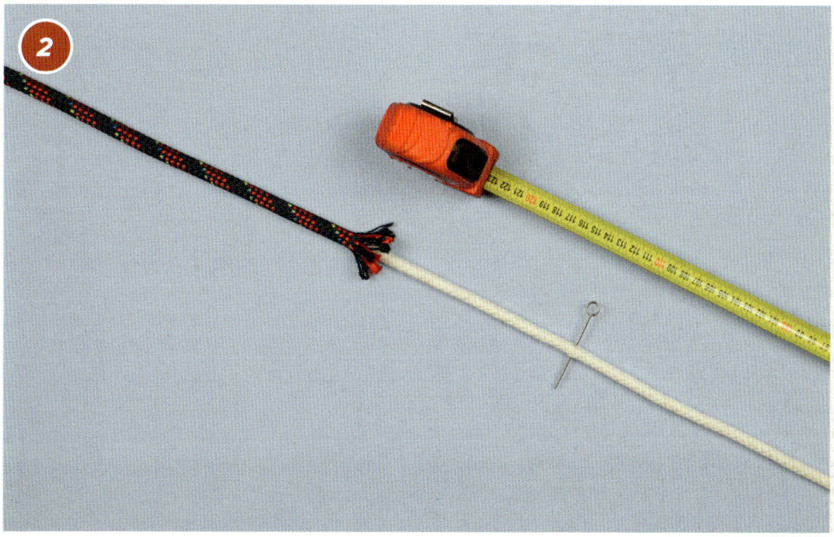

Measure the complete splicing length from the end of the core. It amounts to 48 times the rope's diameter plus the desired circumference of the finished eye splice plus again 48 times the rope's diameter. Here, a 10 mm rope is spliced, so the measurements are 48 cm + 17 cm eye circumference + 48 cm = 113 cm. At this measured distance, push a small needle through the core.

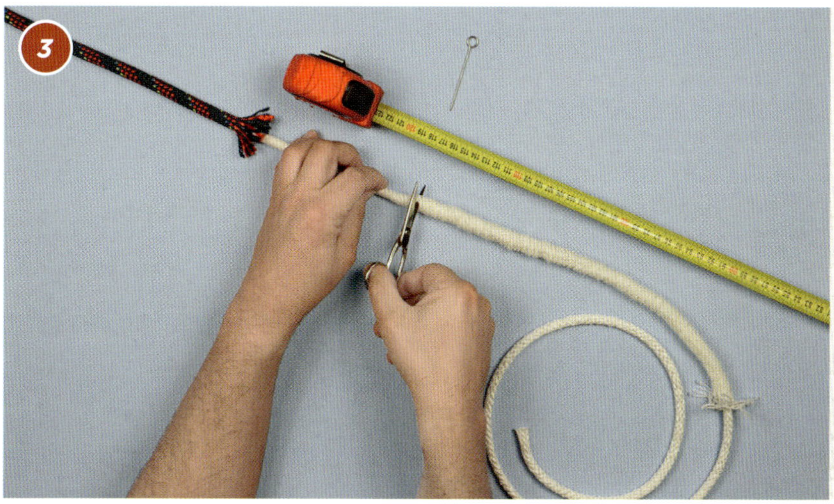

From the end of the core, push back the intersheath until it is bunched against the small needle. Remove the needle and carefully cut the intersheath around the core with a pair of scissors. Make sure not to accidentally cut any fibers of the core.

Slide the cutoff section of the intersheath from the core.

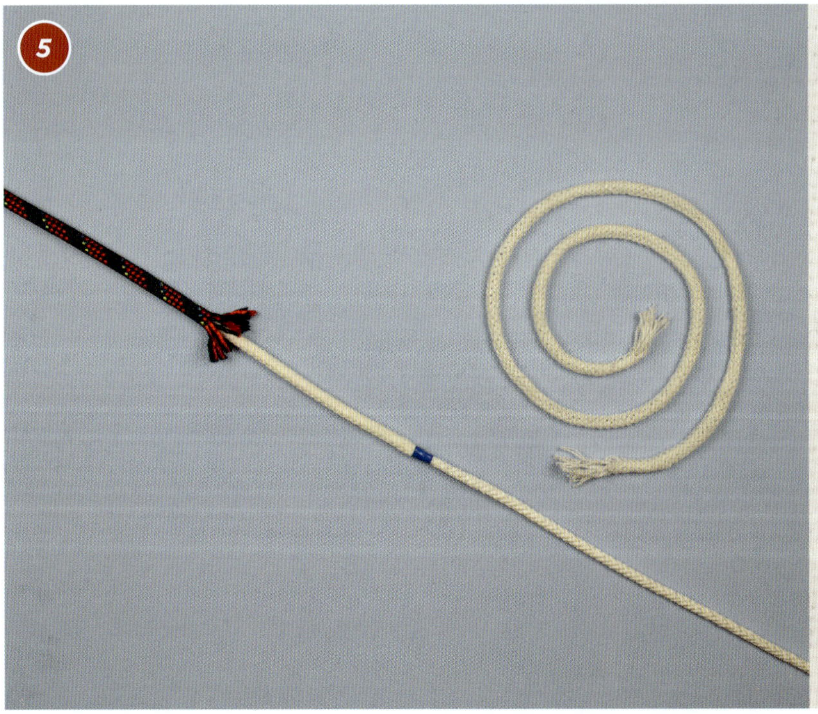

The intersheath has now been removed from the whole splicing length. Tape the end of the intersheath temporarily to the core with a thin wrapping of electrical tape (PVC tape).

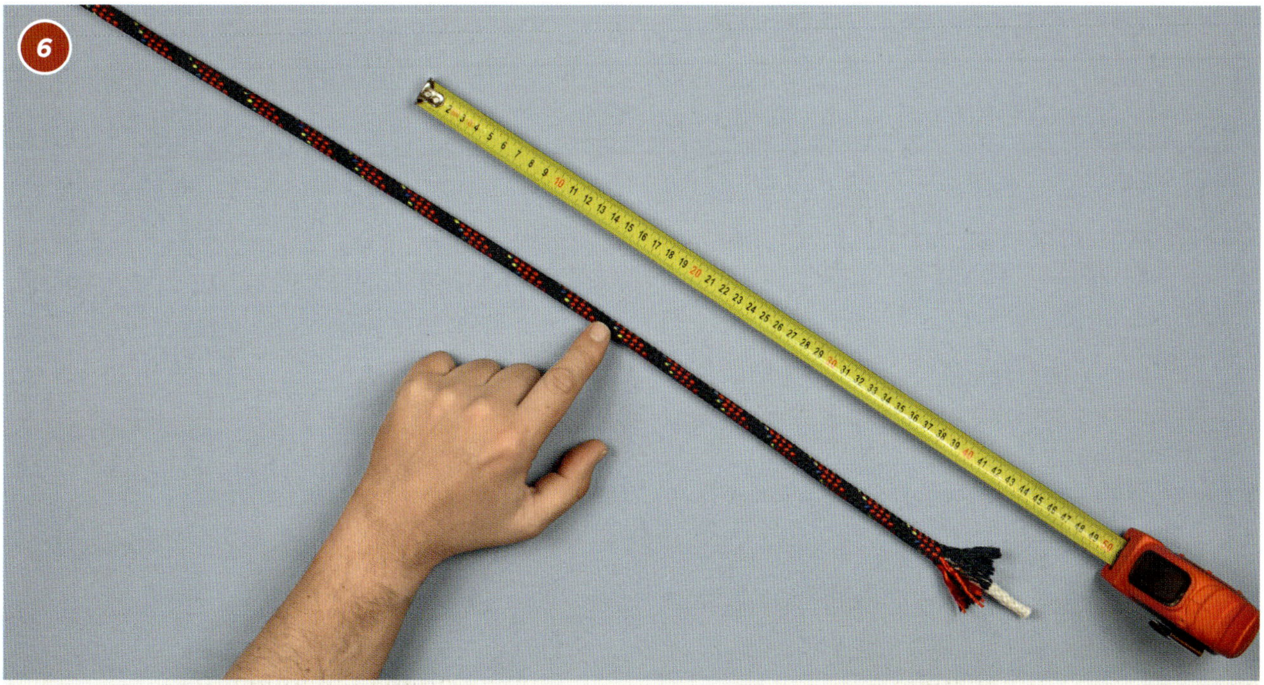

From the butterfly knot, firmly and repeatedly milk back the cover as far as possible. No matter whether the cover end lines up with the core end or not, measure from the end of the core 48 times the rope's diameter; in this case, 48 cm. From this spot the so-called core shortening needs to be measured backward; that is to say, toward the short end of the rope. It has a length of 16 times the rope's diameter. This gives the spot where the core has to be extracted from the cover. How to get the core out of the cover is explained in full detail in the chapter "Basic Splicing Skills" on page 20.

It would have been equally possible to directly measure 32 times the rope's diameter (48 times minus 16 times), but the way that is explained gives a better understanding of the different splicing lengths involved.

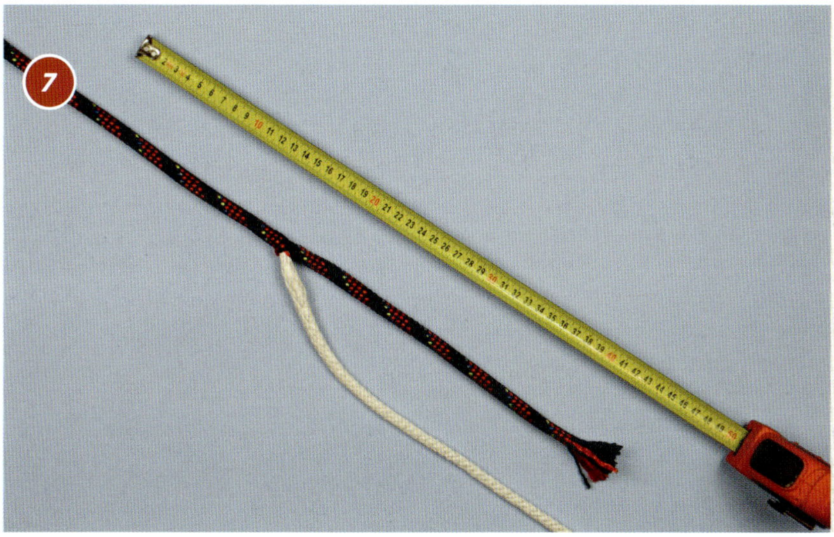

The core has been extracted.

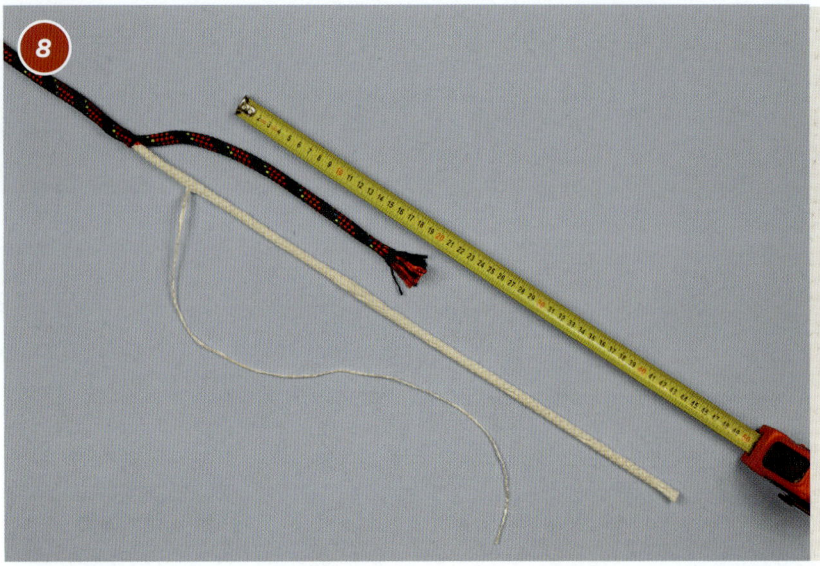

Push the cover back farther and measure 48 times the rope's diameter from the end of the core; here 48 cm. At this spot pull out a single yarn from the core. This is the core marker yarn.

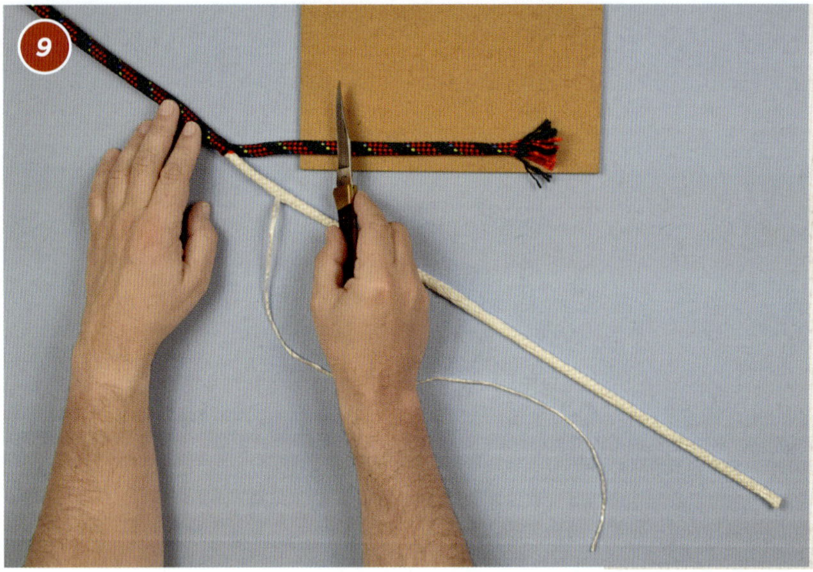

Of the hollow cover end, only a short length of about 12 times the rope's diameter is needed in order to integrate the cover into the splice. Cut the cover end at a distance of about 12 times the rope's diameter from the spot where the core was extracted.

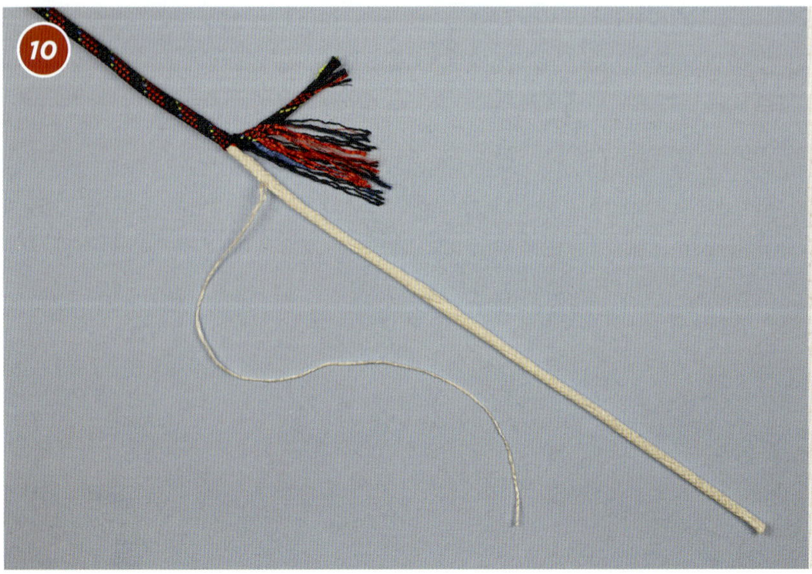

Taper the cover end by pulling out about half of its yarns one behind the other. All details about tapering can be found in the chapter "Basic Splicing Skills" on page 20.

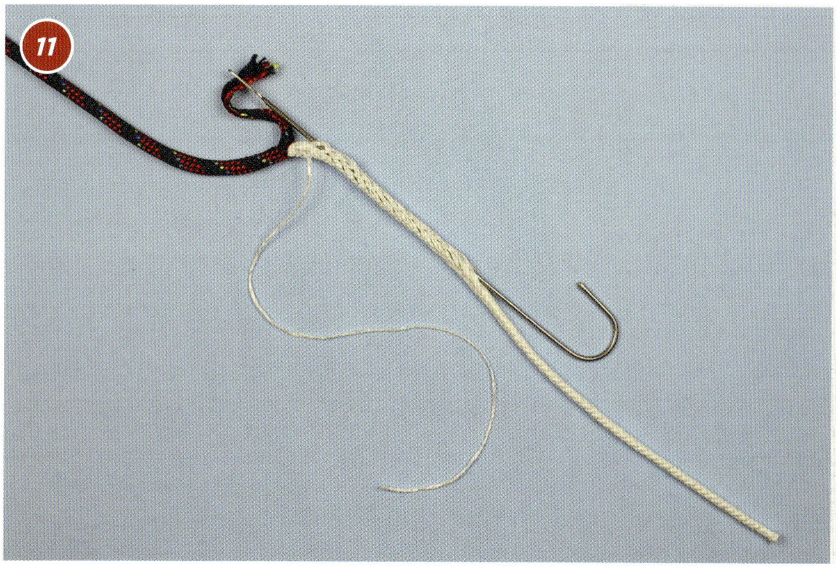

Smooth back the cover from the standing end until the core marker yarn reaches the entry spot. Insert the splicing needle into the core at a sufficient distance and bring it out at the same height as the core marker yarn. Thread the tapered cover end onto the splicing needle and pull it through.

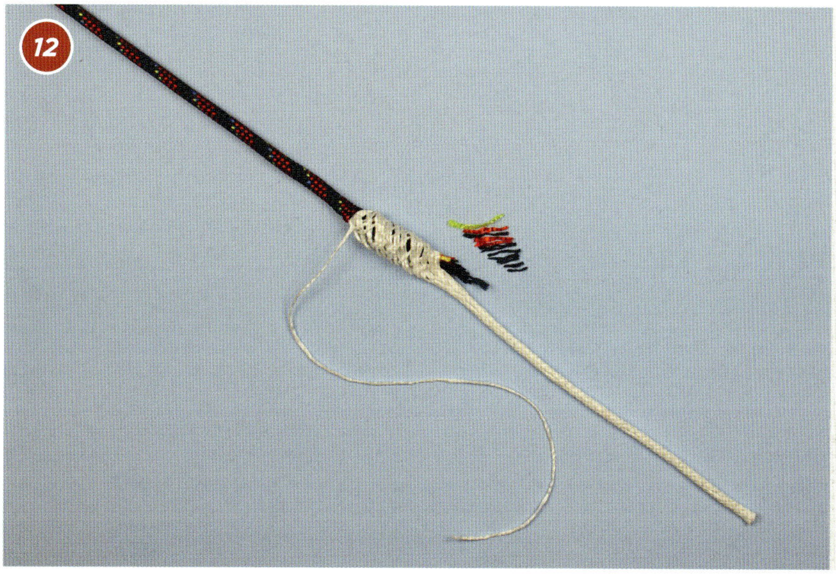

Firmly pull at the cover end while pushing back the core and then firmly milk back the cover from the standing end of the rope. This pushes the cross-over spot of core and cover tightly together. Now cut the protruding cover end at a sharp angle and smooth back the core. The cover end disappears inside the core.

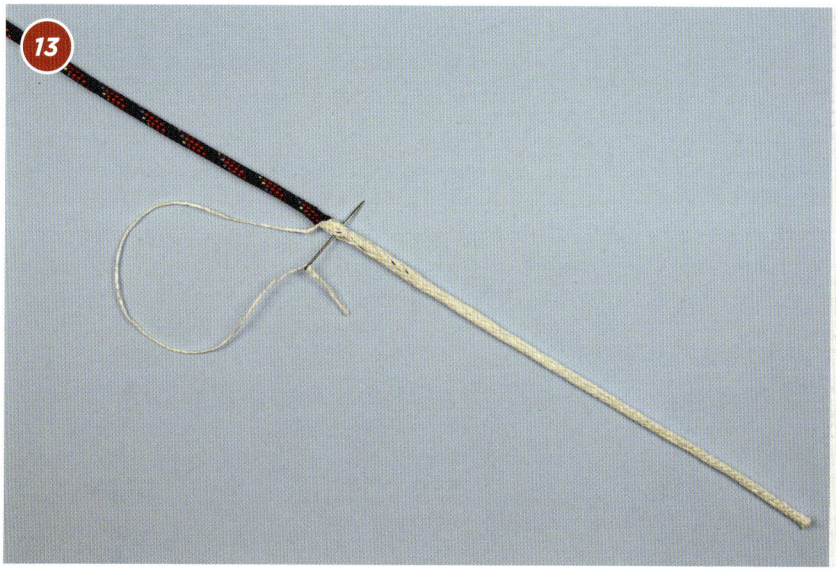

Firmly smooth and milk a few more times from the crossover toward the end of the core until the exposed core lies tightly compressed around the cover end. Then secure the cross-over with the core marker yarn by making about five closely spaced stitches toward the side of the exposed core. Afterward the core marker yarn is cut off flush.

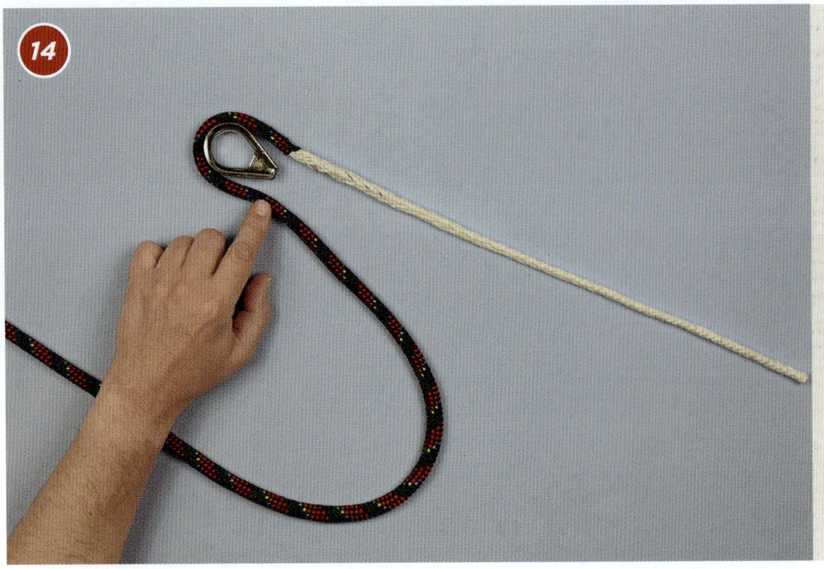

Form the eye in the desired size. Here, a thimble is to be spliced into the eye. The eye circumference starts at the crossover spot. At the spot where it ends (pointed out in the picture), the core needs to be extracted a second time from the cover so that it forms a big loop.

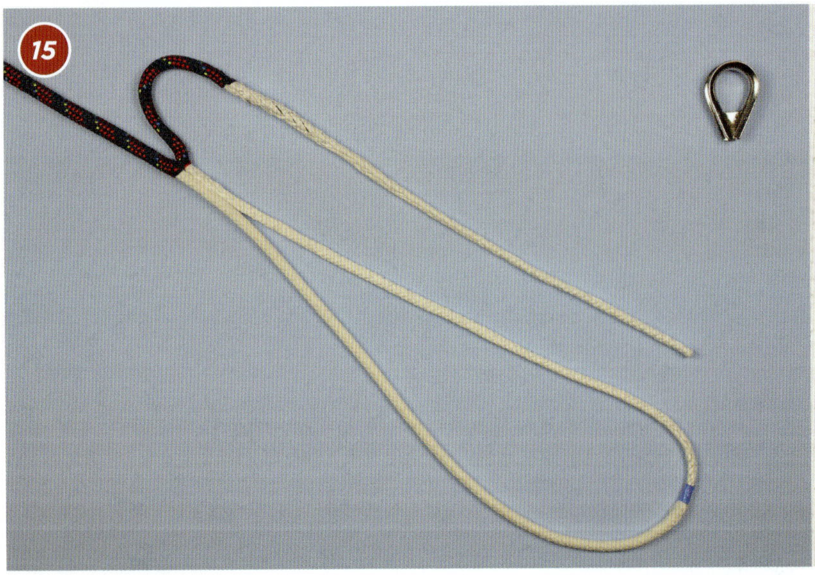

The core has been extracted and pulled out from the side of the standing end so that it forms a big loop. The end of the intersheath that was taped to the core also reappears.

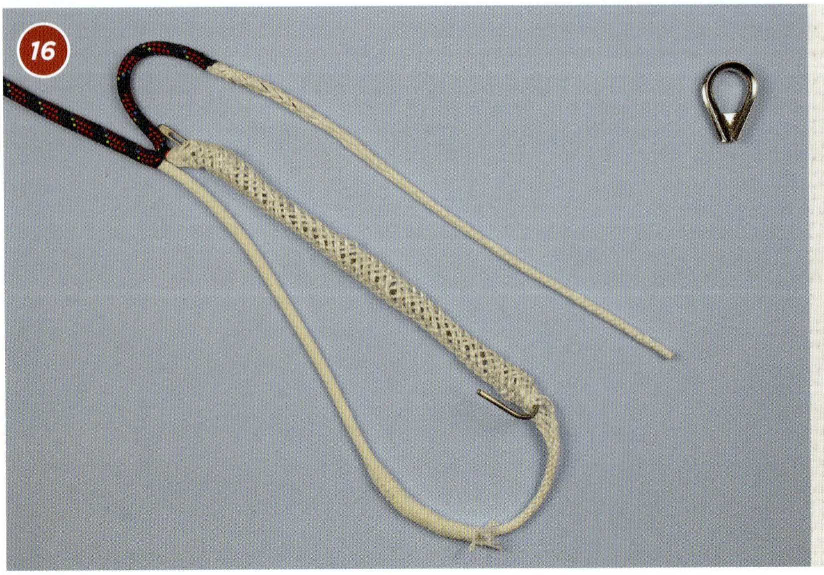

Remove the tape and push back the intersheath a short distance. Insert the splicing needle at the spot where the intersheath was taped to the core, and bring it out at exactly the spot where the eye circumference ends. Make sure the cover is not bunched in the section of the eye circumference; smooth back from the crossover spot if necessary.

Thread the core end onto the splicing needle and pull it through.

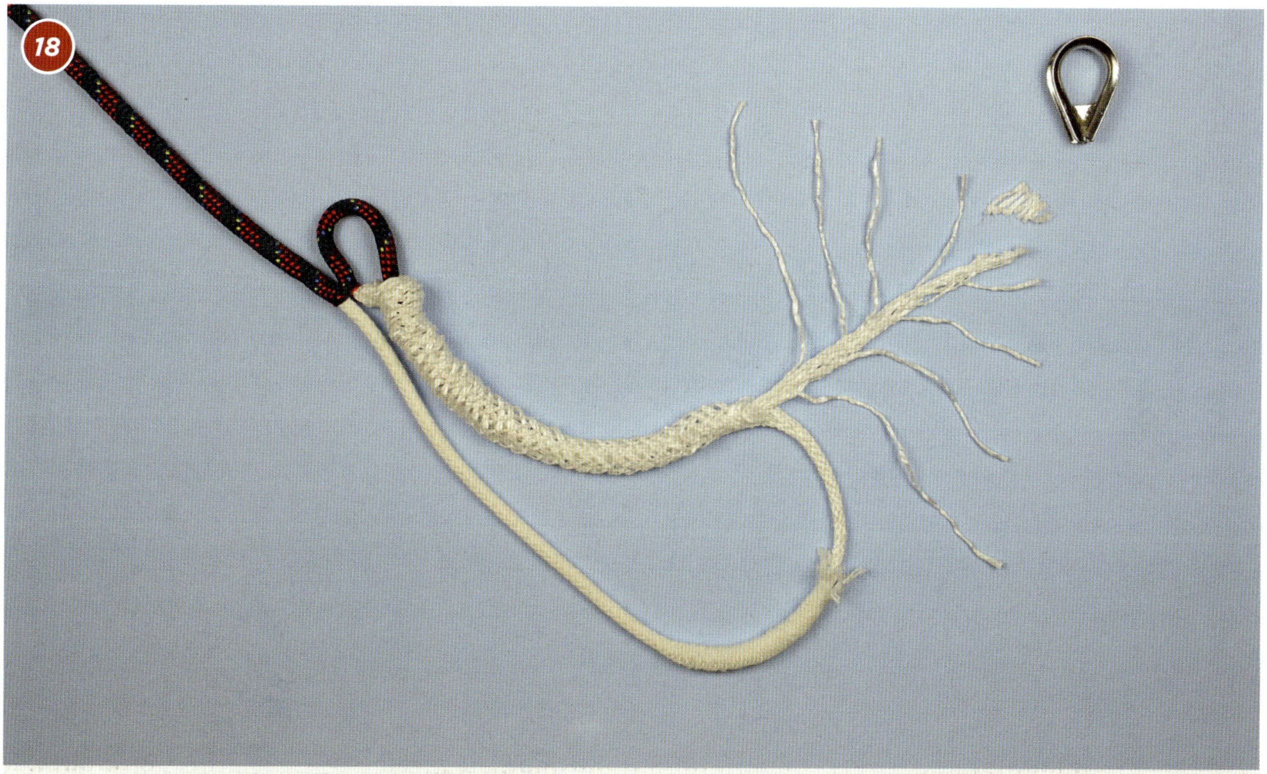

Taper the core end over the last third of its length by carefully pulling out about half of its yarns one behind the other. Then cut the very end of the core at a sharp angle.

TIP: *Some ropes require the core to be tapered right from the crossover spot and not only over the last third of the core end, in order to be able to later milk the thickened spliced section into the core. In this case, taper the core end before pulling it through. But this is an exception and is necessary only with very few ropes that have an extremely tightly braided cover.*

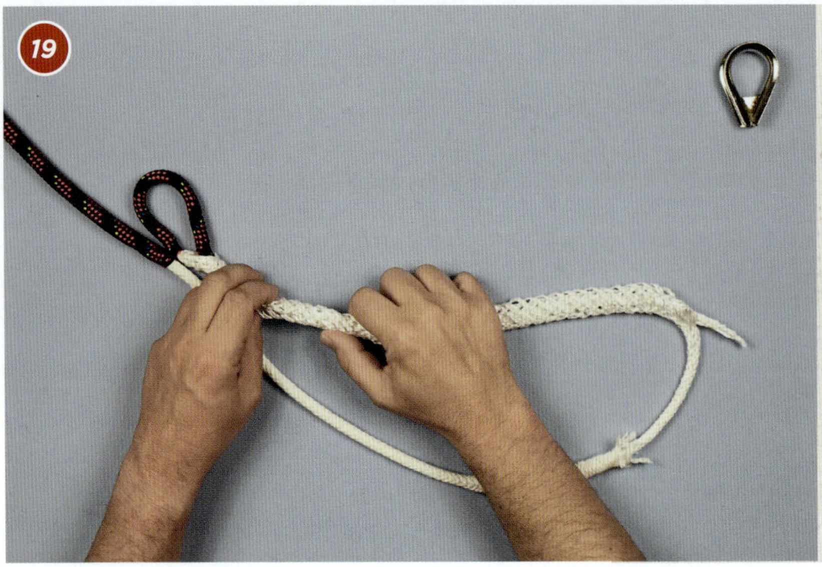

Firmly smooth back over the core. The core end disappears inside. Continue smoothing back over the core until it is lying nicely compressed over the core end inside.

The end of the core has disappeared inside. Smooth the bunched intersheath back toward the spliced section. The end of the intersheath does not need to be affixed to the core again.

If you have worked very precisely, the end of the intersheath now overlaps with the end of the tapered core on the inside by a few centimeters and gives a perfect transition. In case this did not work out perfectly, it is of only aesthetical concern, however, and has almost no influence on breaking strength.

It is advisable to temporarily secure the spot where the core enters into the core. Push a small needle across this spot or tie a short length of whipping twine around it, making a constrictor knot.

Now everything is ready for milking back. Attach the rope with the butterfly knot to a very strong point. If working at home, do not attach to door handles or the like. They are not strong enough! Now we'll need a little pulling force! All details about milking back, including tips and tricks, can be found in the chapter "Basic Splicing Skills" on page 20.

Up to this point it is fairly easy. Always keep the core tensioned as it is drawn inside the cover.

The thimble is inserted at the last moment, when the eye is just still big enough. The last few centimeters or inches demand more pulling force. Use a hook or handle, as shown, to be able to really pull at the end. If pulling by hand alone is not enough, put the eye splice under more tension by using a purchase or a winch. Try to avoid excessive bunching of the core at the entry point. A little bit is okay and will later straighten out completely.

Done! This is what it should like at the end: the splice is closed completely and is very tight around the thimble.

Once again, the most-important steps of making an eye splice, shown with a selection of other first-grade ropes of different manufacturers (all double braided with a core from Dyneema).

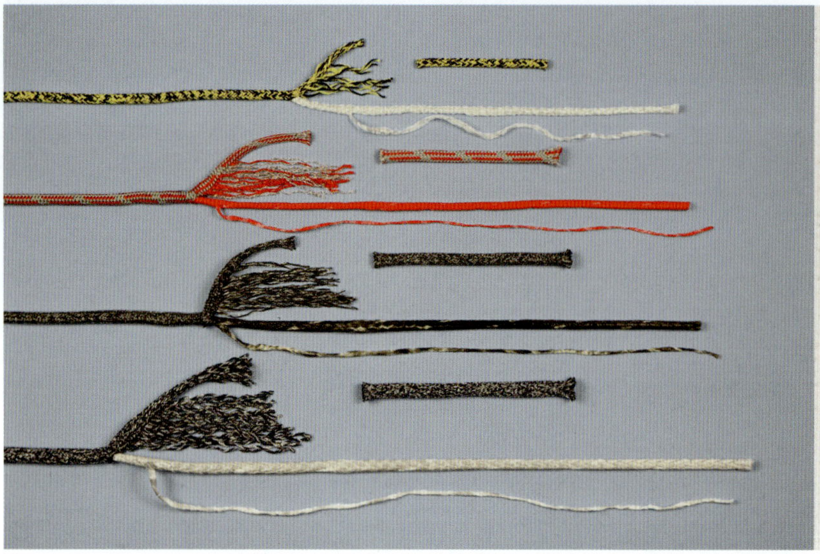

Tapering the cover end; compare step 10.

The core end with the integrated cover end; compare step 13.

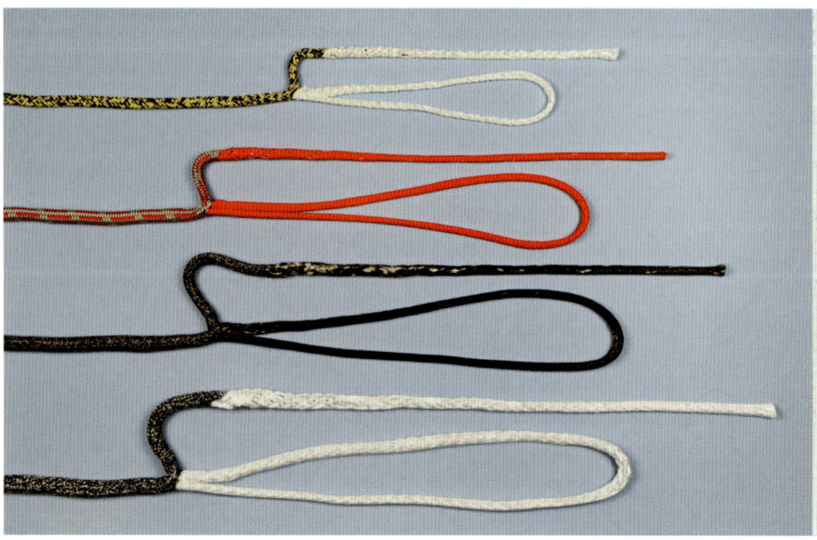

The core has been extracted a second time and forms a big loop; compare step 15.

NOT SO GOOD:

Sometimes you may see an eye splice that has the cover end either simply removed at the eye (*right*) or just tucked through the eye and pulled back into the rope (*left*). With these versions, the load-bearing point inside the eye and the eye as a whole remain completely unprotected! In principle this splice resembles a splice in single-braided rope, with a cover just pulled up over it.

2.5 END-TO-END SPLICE IN HIGH-MODULUS FIBER ROPE

Double braided ropes (kernmantle ropes) with a core from high-modulus fiber such as Dyneema, Spectra, Vectran, and other high-tech fibers.

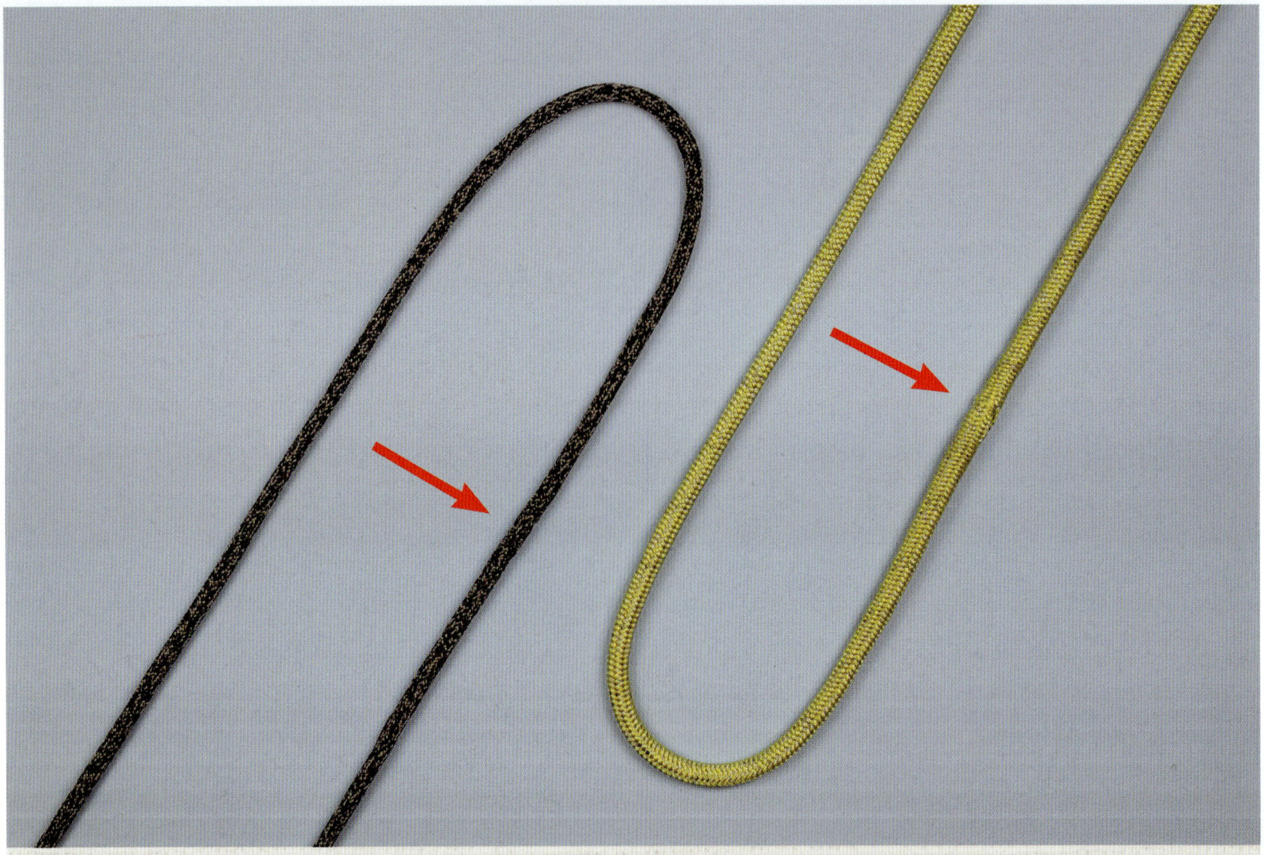

Ropes with a core of high-modulus fiber need to have the core spliced to the core to achieve maximum breaking strength in an end-to-end splice. But quite often it is more important to avoid any increase in diameter at the splice than to maximize breaking strength. Whether the end-to-end splice results in an increase in diameter or not depends on the amount of tapering of the ends. A well-made end-to-end splice is hardly visible, and it needs a close look to discern the spot where the ends of a line have been joined.

1. Method: End-to-end splice with the core spliced to the core

This method ensures high breaking strength. However, there will be a slight increase in diameter at the splice. With carefully executed additional tapering, it is possible to almost completely avoid this increase in diameter.

START

VIDEO 2.5

Construction:	double braided (kernmantle rope) with or without intersheath
Material:	core: high-modulus fiber (e.g., Dyneema, Spectra), cover: polyester
Difficulty:	a little more challenging
Splicing tools:	splicing needle or fid

1

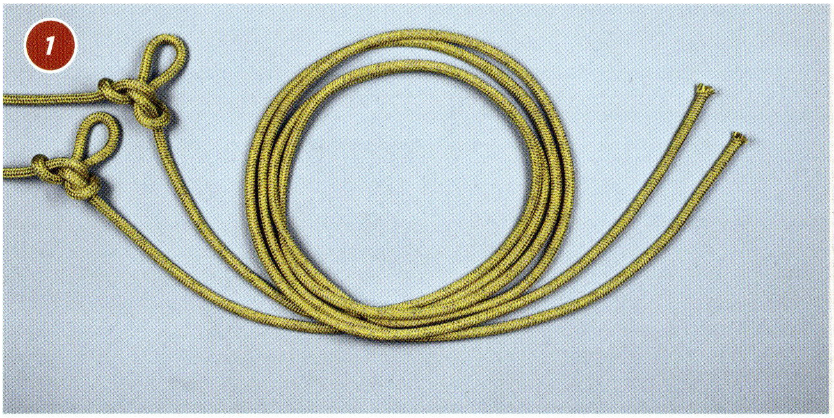

Tie a loop knot, preferably a butterfly knot, at about 3 meters / 10 ft. from each end of the rope. The ends need to be cut with a knife and must not be fused.

2

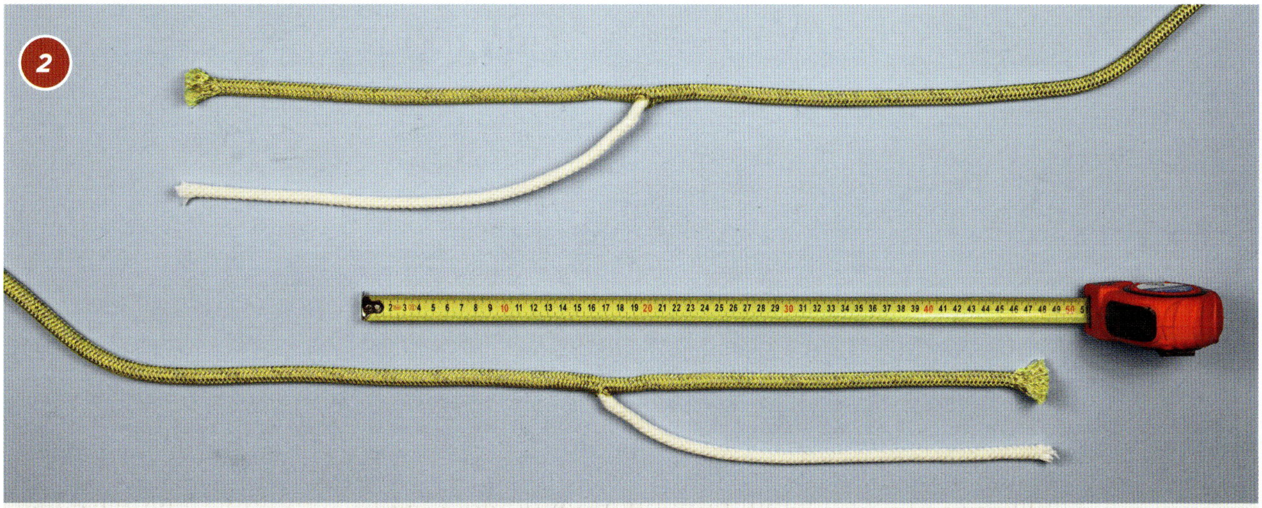

From now on, all steps performed on one end of the rope are to be replicated in exactly the same way on the other end of the rope. Measure a distance of 48 times the rope's diameter from the end of the line; here, 48 x 10 mm diameter = 480 mm or 48 cm. From this spot, the so-called core shortening needs to be measured backward; that is to say, toward the short end of the rope. It has a length of 10–16 times the rope's diameter (on the short side if a stronger taper is to be executed at step 11, and on the long side if no intersheath is present). This gives the spot where the core is extracted from the cover.

In case the rope has an intersheath, extract the core together with the intersheath. How to get the core out of the cover is explained in full detail in the chapter "Basic Splicing Skills" on page 20.

(To avoid any increase in diameter at the splice, as far as possible, the taper at step 11 needs to be more pronounced, as is described there. In this case, reduce the length of the core shortening down to 10 times the rope's diameter.)

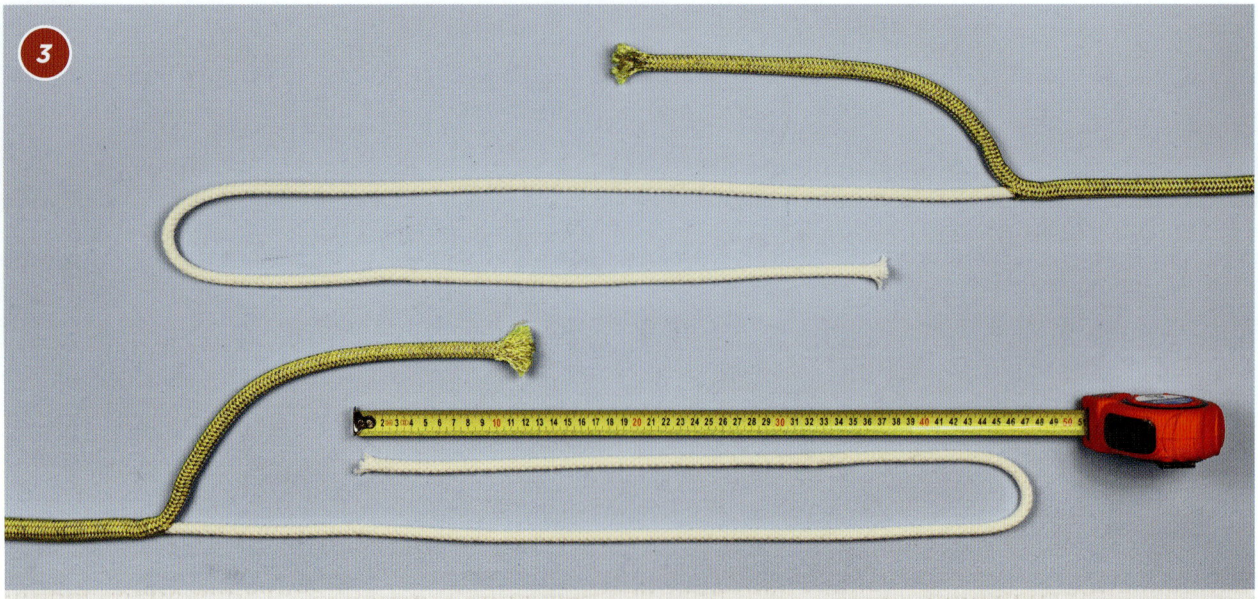

Pull out more of the core (including the intersheath if present) by pushing back the cover. Now a length of twice the 48 times of the rope's diameter needs to be measured from the end of the core (here, 2 x 48 cm). To facilitate measuring, the core may be laid out double, as shown in the picture. The spot to be measured is opposite the end of the core.

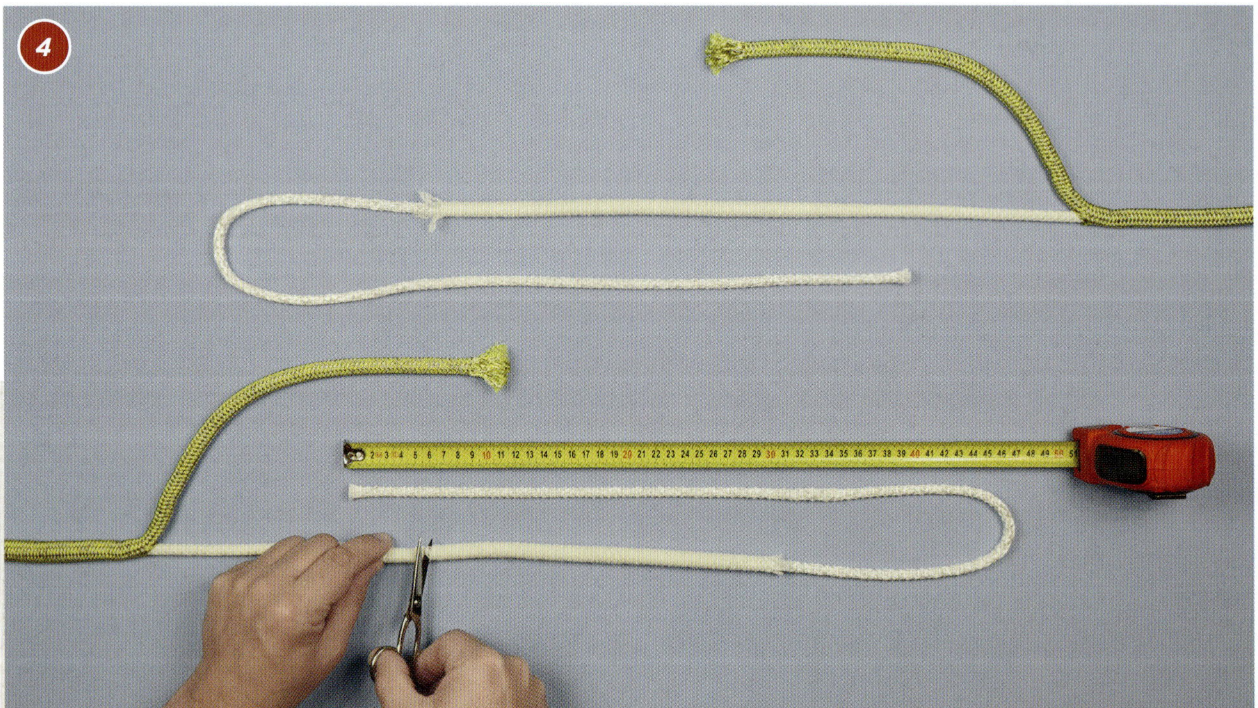

If the rope has an intersheath, it is now removed from the splicing length. Push back the intersheath from the end of the core so that it is bunched and detached from the core. Cut the intersheath around the core with a pair of scissors a few centimeters ahead from the measured spot opposite the core end, as shown.

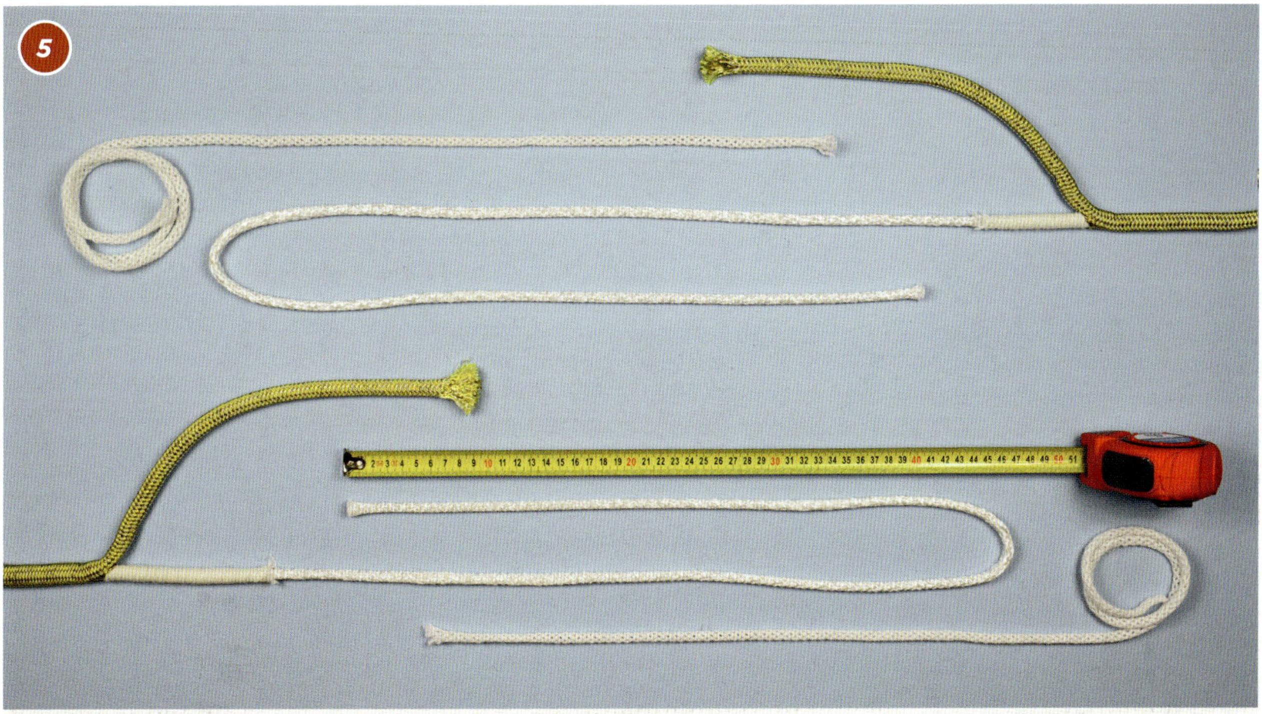

Slide the cutoff section of the intersheath from the core.

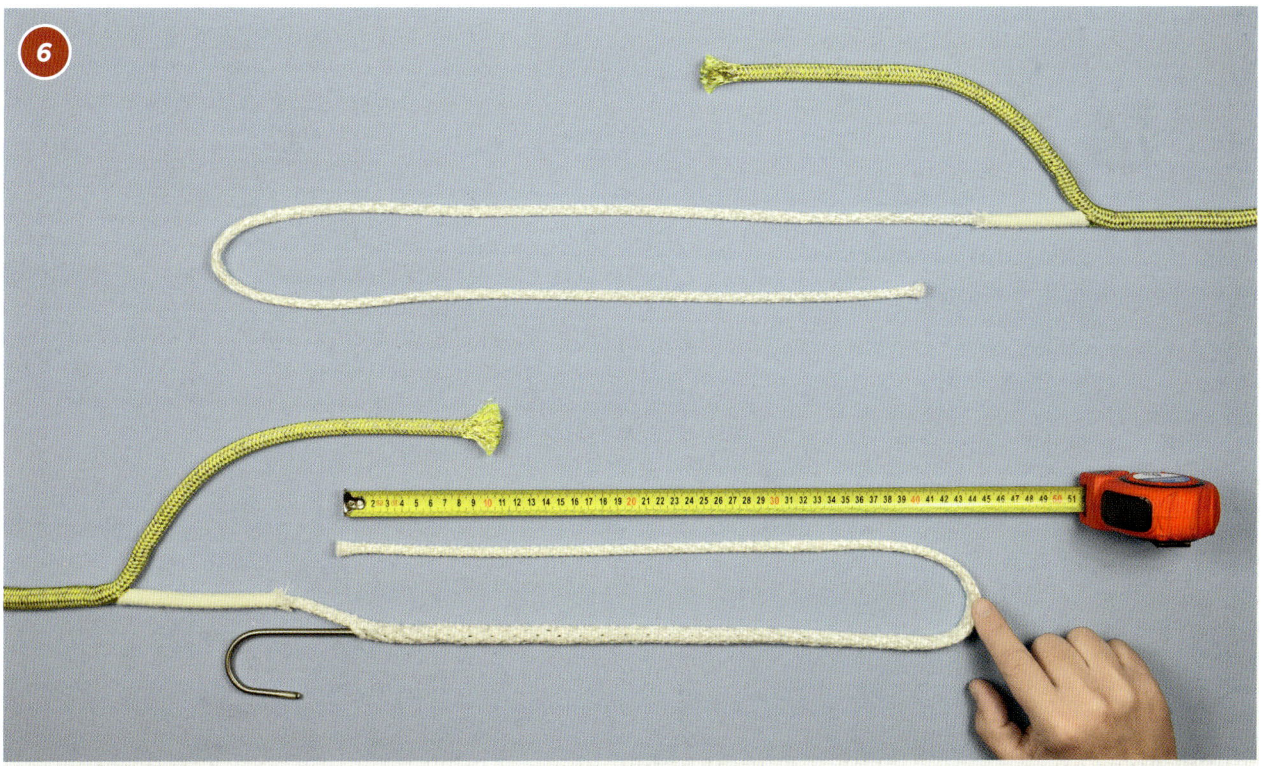

Insert the splicing needle at the measured spot and bring it out at half the distance from the core end (pointed out).

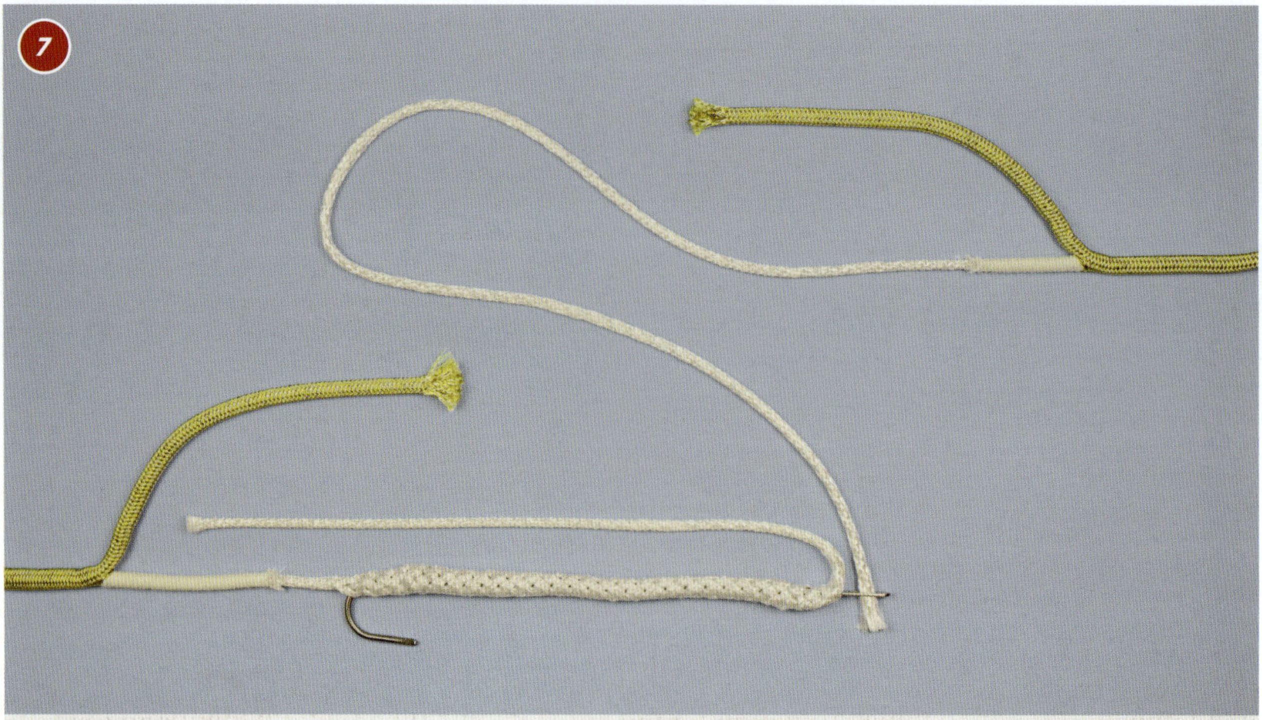

Thread the other core end onto the splicing needle and pull it through.

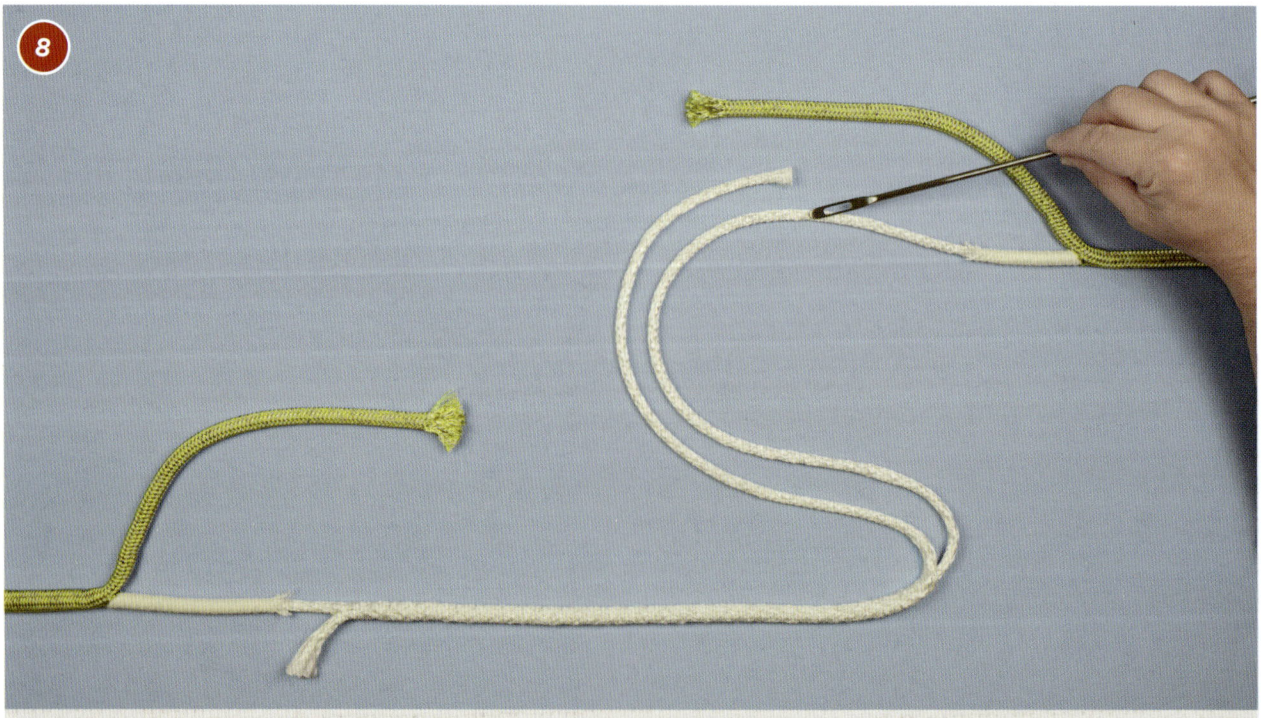

The end of the core that was pulled through protrudes only a few centimeters. Now insert the splicing needle into the other core end, as shown, and bring it out right next to the spot where the one core enters into the other core end. Pull the other end through.

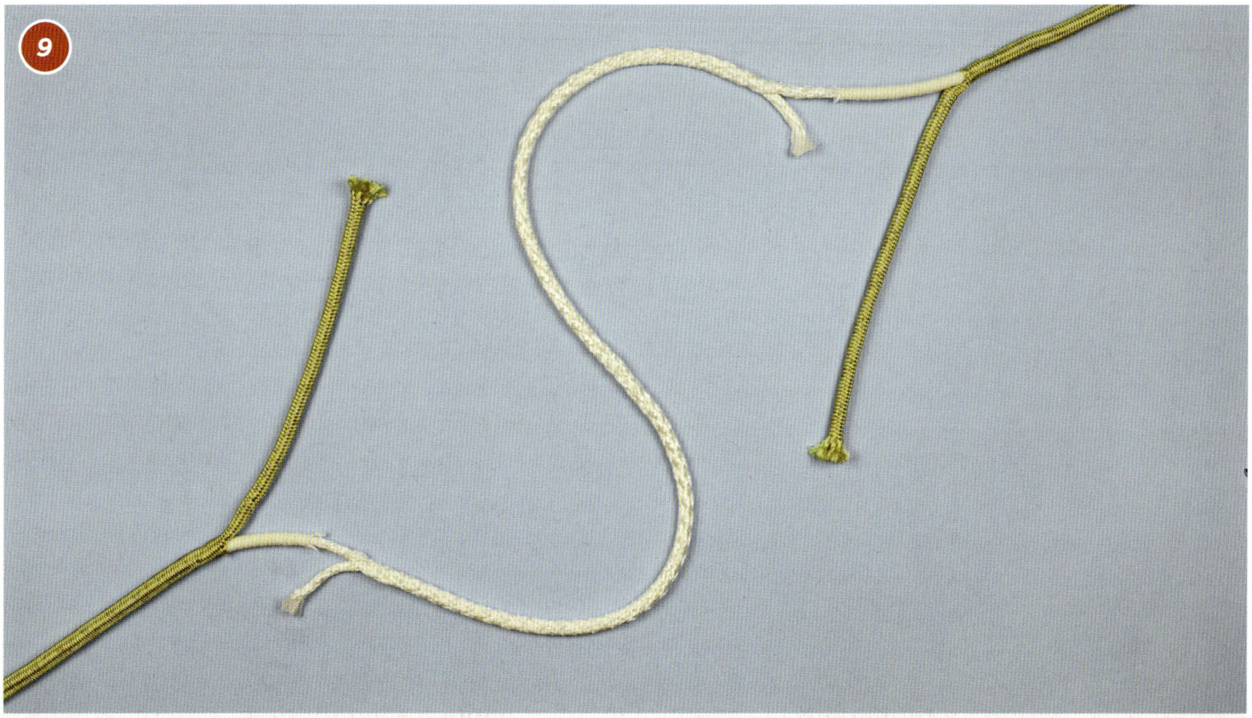

Both core ends have been spliced into their opposing ends. The ends protrude by a few centimeters.

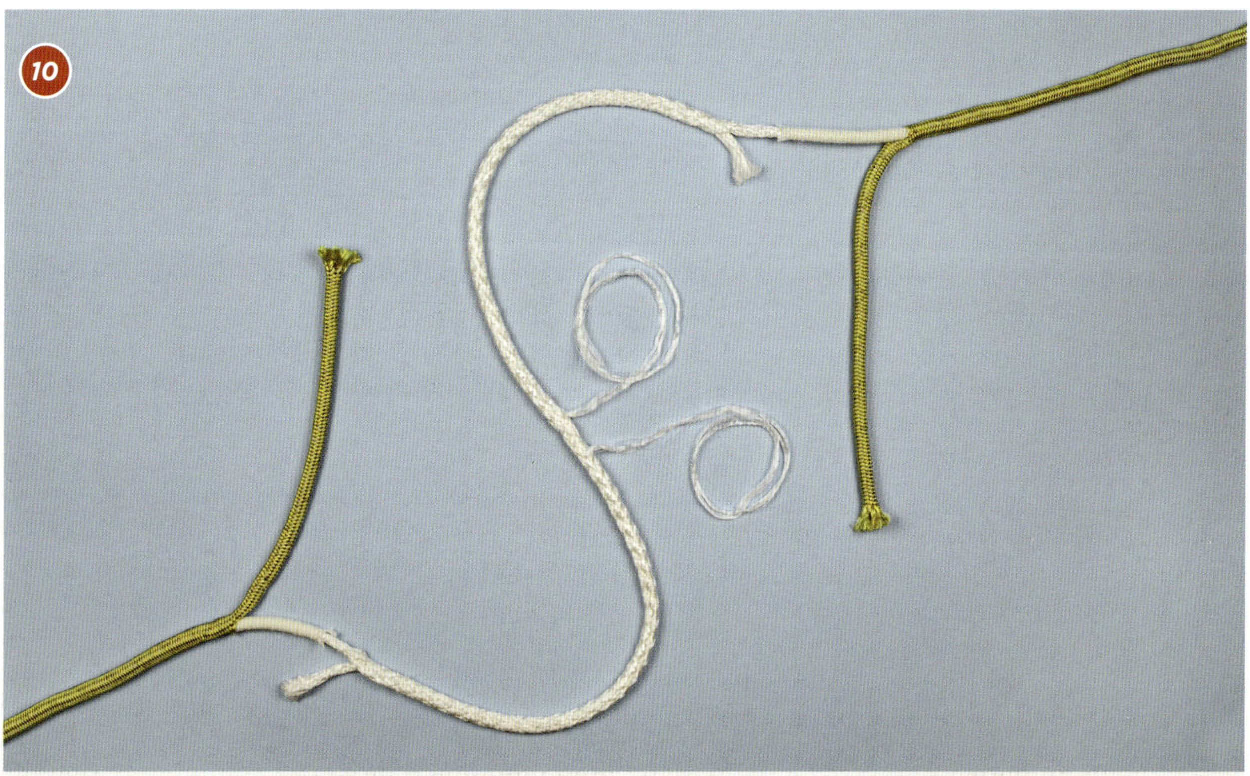

At each side of the crossover, pull out a single yarn from the cores.

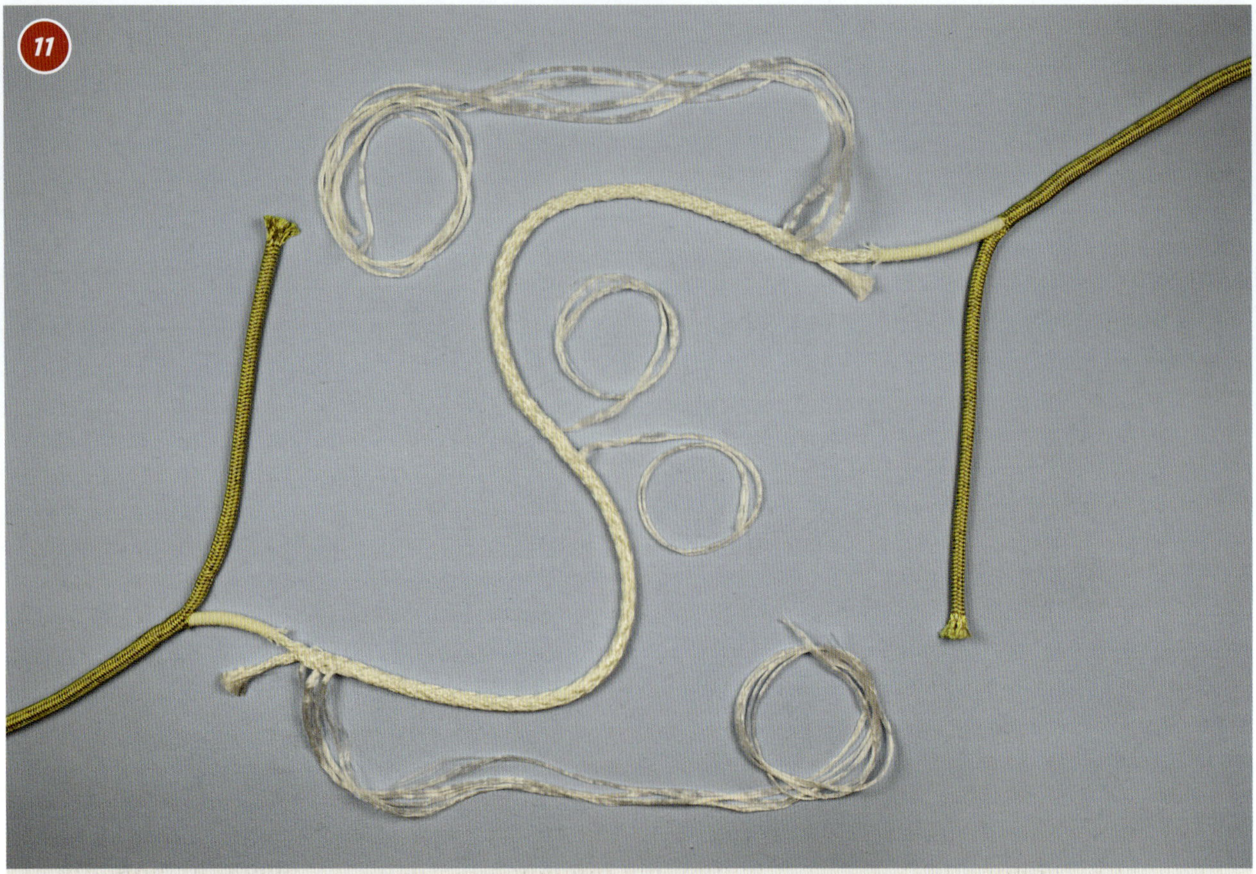

Now the cores are tapered. To minimize any increase in diameter at the finished splice, start the taper right at the exit spots of the core ends. Work extra carefully when pulling out the yarns for the taper over such a great length. Hold down the cores onto the work surface, so that the cores do not bunch too strongly when pulling out the yarns one behind the other. The more yarns are pulled out, the less the increase in diameter at the finished splice. If half the yarns of the core are pulled out, there will hardly be any increase in diameter. To not compromise the braid of the cores, working very carefully is of utmost importance!

(If by chance you catch and pull out one of the yarns that were already pulled out next to the crossover, you will have to pull out another yarn at the crossover again.)

The yarns pulled out for the taper have been cut off flush. Secure the crossover with the two yarns pulled out next to it by making a few stitches toward both sides of the crossover.

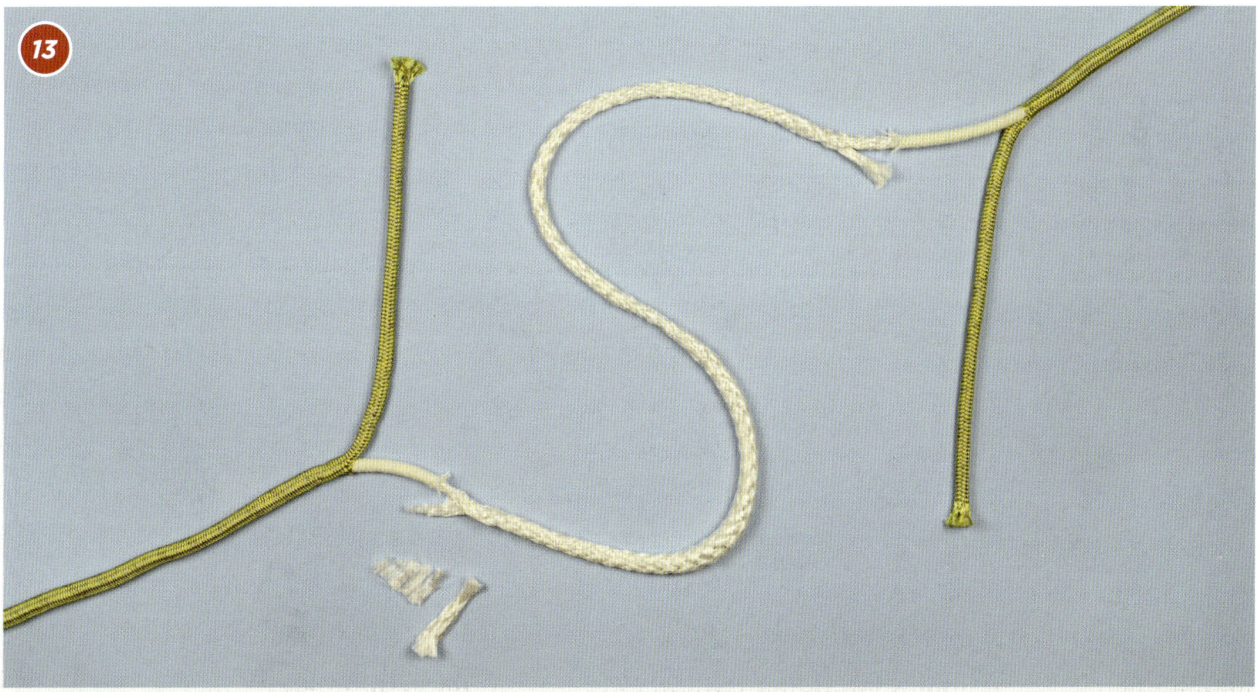

Now taper the ends of the cores. Hold the end at exactly the spot where it emerges, and pull it out a little farther. Cut it off at the spot where you are holding it. Then open up the braid over the last few centimeters and cut the very end at a sharp angle. Smooth back the outer braid and the end disappears inside, giving a nice transition.

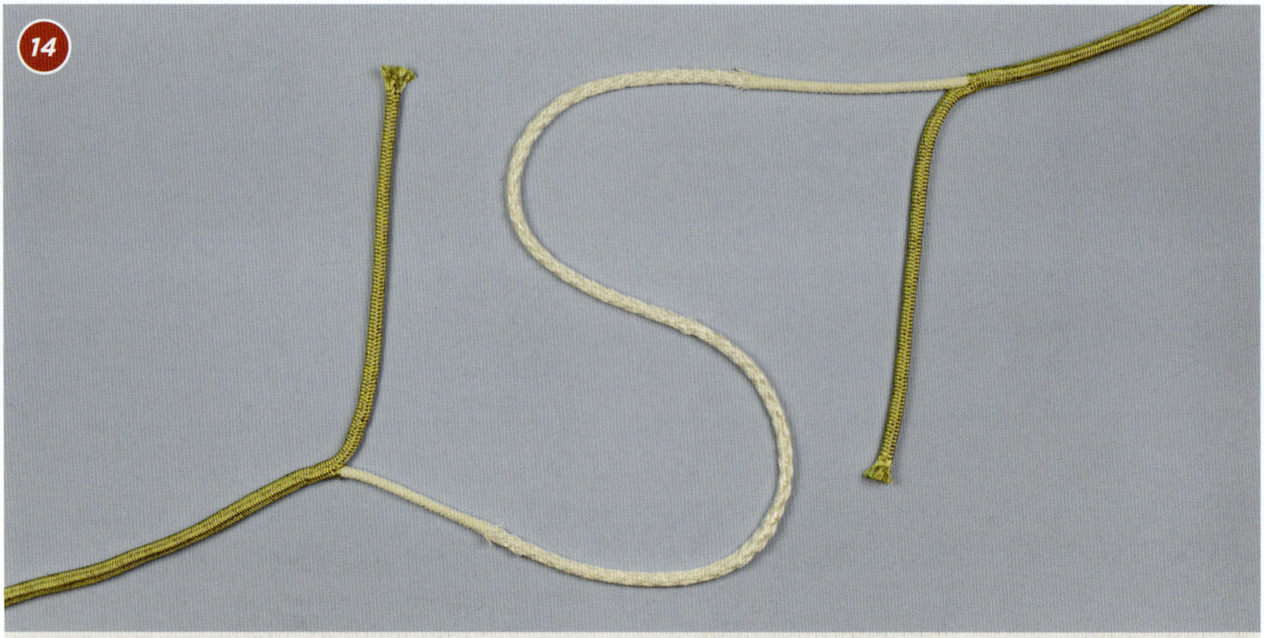

From the other direction, smooth back the intersheath. It should just overlap the spot with the core end inside. If the overlap stretches farther, shorten the intersheath appropriately.

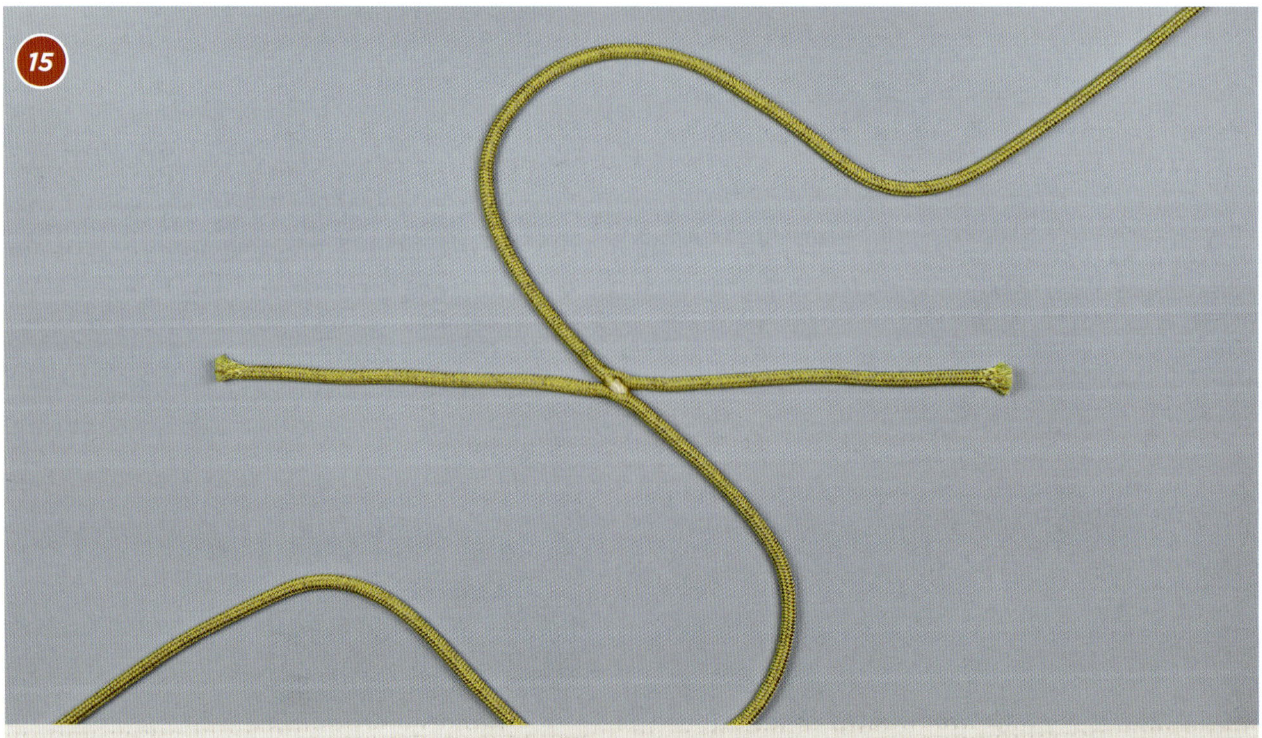

Alternately attach the rope with one of the two butterfly knots to a strong point and start milking back until the cover completely closes over the spliced cores. All details about milking back can be found in the chapter "Basic Splicing Skills" on page 20.

Make sure to avoid any torsion (twist) in the covers when they close over the spliced cores. Then cut the cover ends off so that a length of only about 12 times the rope's diameter remains. Taper these ends, starting at about one rope diameter away from the entry spot of the cores. It's best to pull out the yarns for the taper on the side oriented toward the rope.

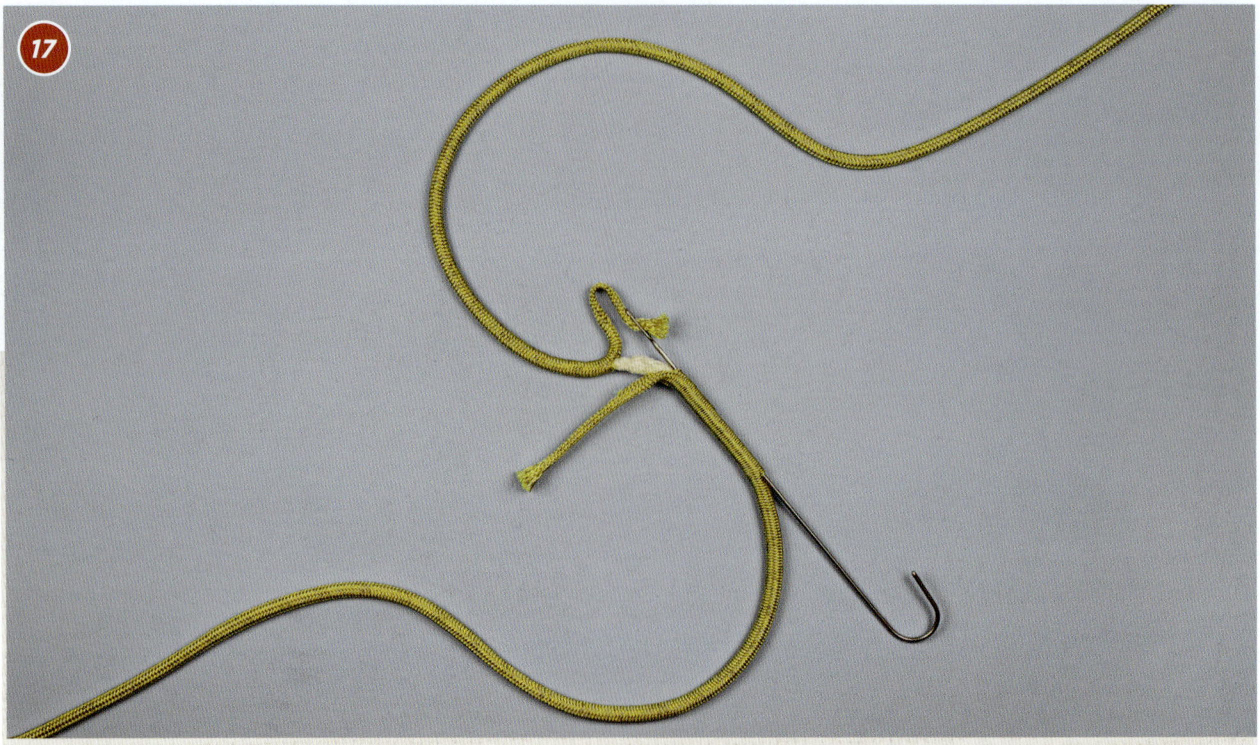

The tapered cover ends are pulled inside with the splicing needle. Important: Make sure to insert the splicing needle between the spliced core and the cover. Do not pierce into the core!

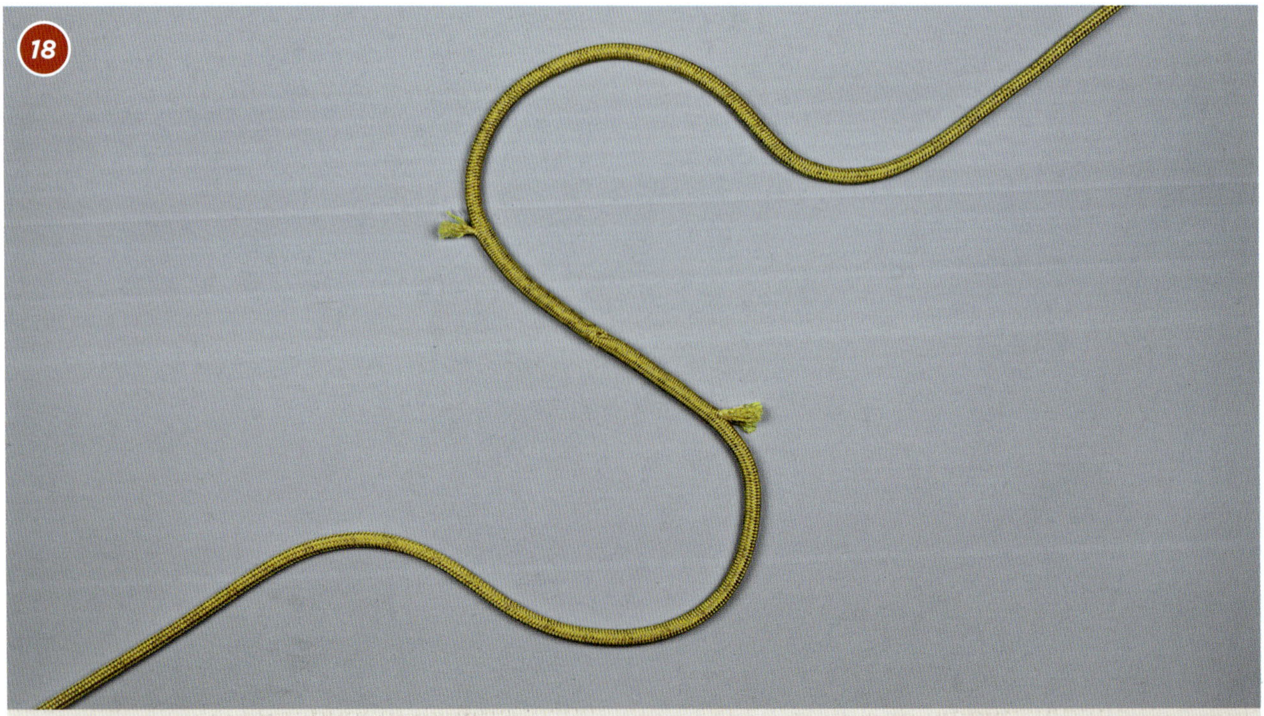

Pull on the protruding cover ends to tightly close the crossover of the cover.

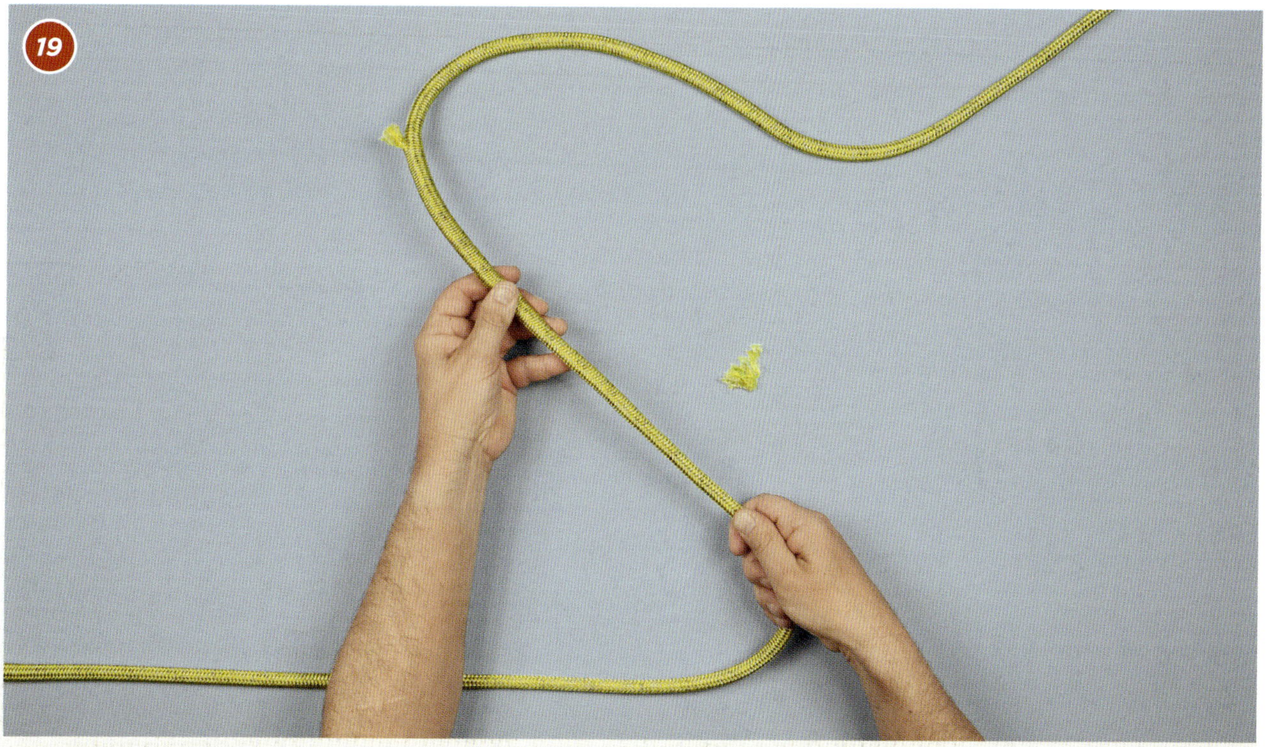

Cut the protruding cover ends at a sharp angle and milk back from the crossover. The cover ends disappear inside.

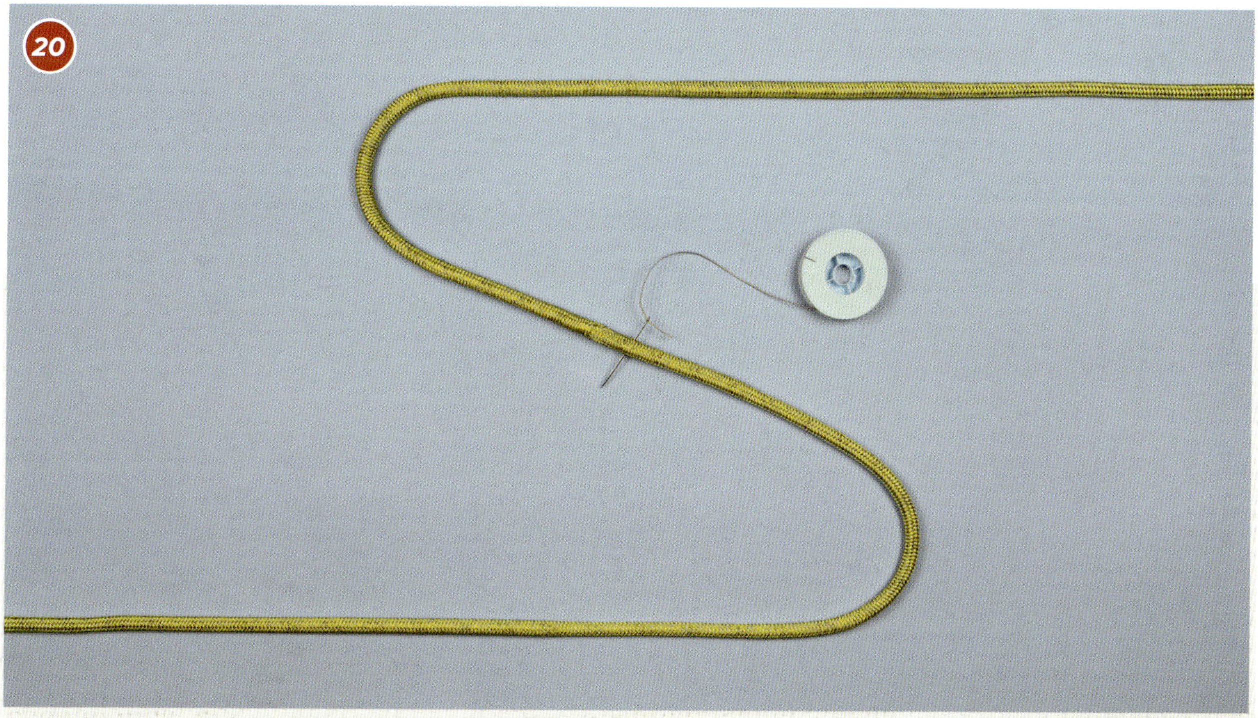

Use a short length of whipping twine to secure the crossover with a few stitches.

Method: End-to-end splice with the core spliced to the core

With this method, the cover ends are spliced into each other. The core ends are cut off but overlap the cover ends on the inside. There is practically no increase in diameter at the splice. This splicing technique can in principle also be used for ropes with a core from conventional fiber; compare chapter 2.3.

START VIDEO 2.5.1

Construction:	double braided (kernmantle ropes) with or without intersheath
Material:	core: high-modulus fiber (e.g., Dyneema, Spectra), cover: polyester
Difficulty:	medium to challenging
Splicing tools:	splicing needle or similar pull-through tool

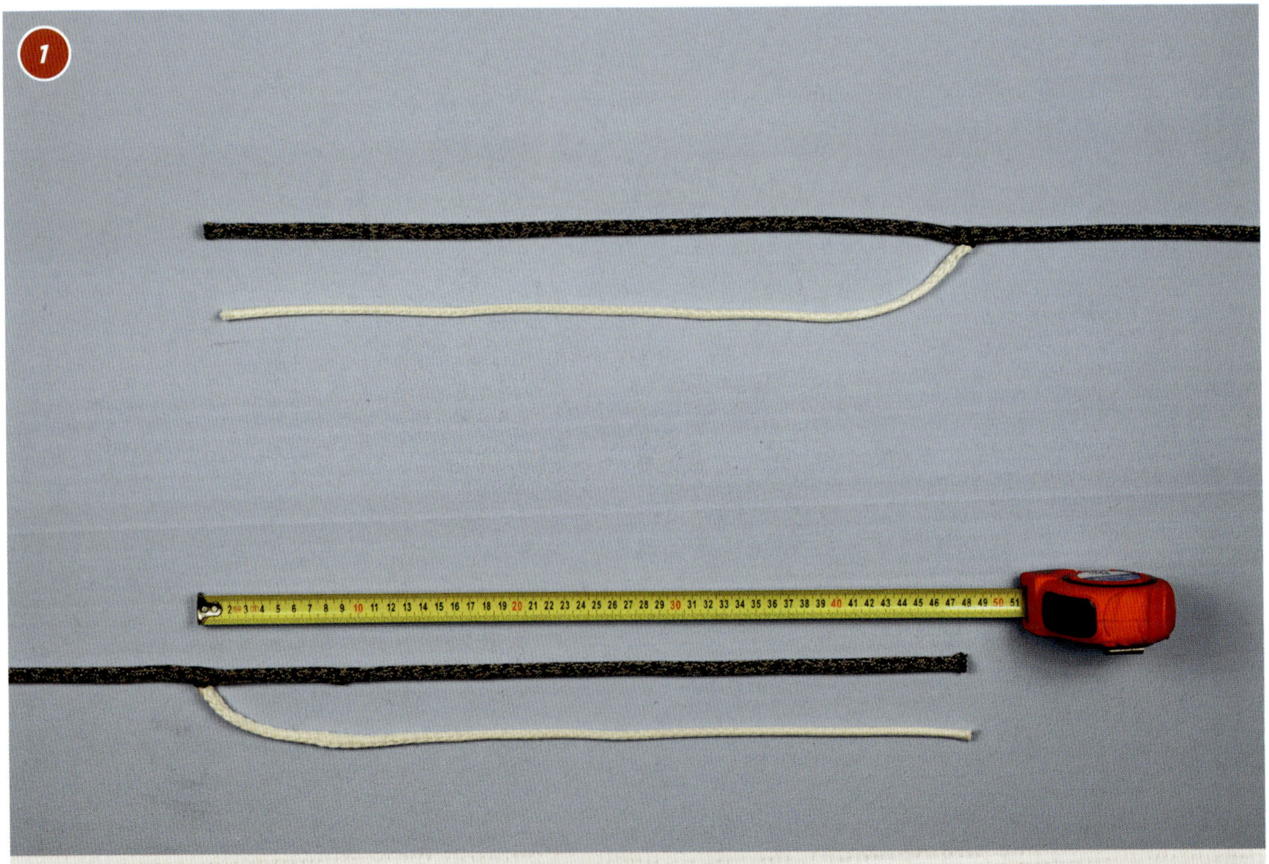

Measure 60 times the rope's diameter from the end of the rope and extract the core at this spot. Here, 60 x 8 mm diameter = 480 mm or 48 cm.

If an intersheath is present in the rope, it needs to be removed over a length of 80 times the rope's diameter before the core is extracted. How an intersheath is removed is detailed in chapter 2.4, steps 1–5.

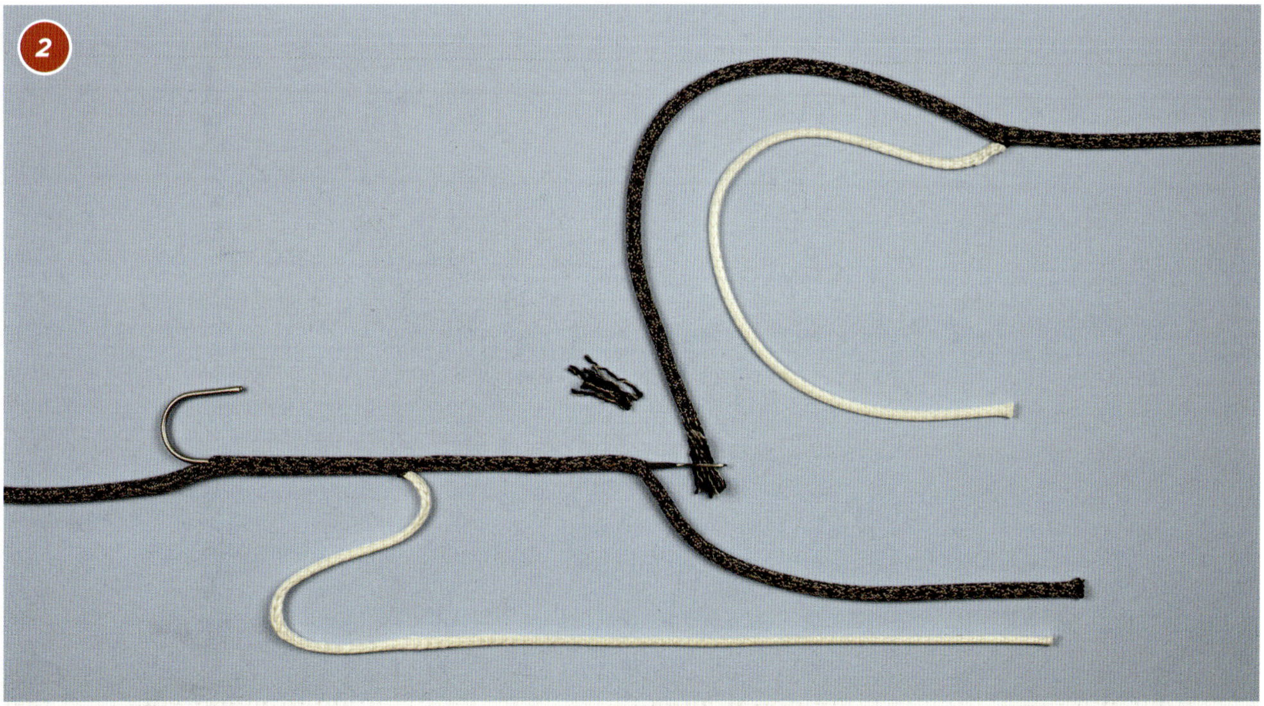

From the spot where the core emerges, measure 20 times the rope's diameter toward the standing end and insert the splicing needle at this spot, as shown. Push the splicing needle along between the core and the cover. At a distance of, again, 20 times the rope's diameter from the spot where the core emerges, but this time toward the short end, is where to exit with the splicing needle. That means in total the splicing needle is pushed along inside the rope over a distance of 40 times the rope's diameter. Now thread the other cover end onto the splicing needle. It is best to open up the braid at the very end and thread only about half the yarns of the cover braid onto the splicing needle. The yarns that do not get threaded onto the splicing needle are carefully cut off. That way, it is easier to pull the cover end through.

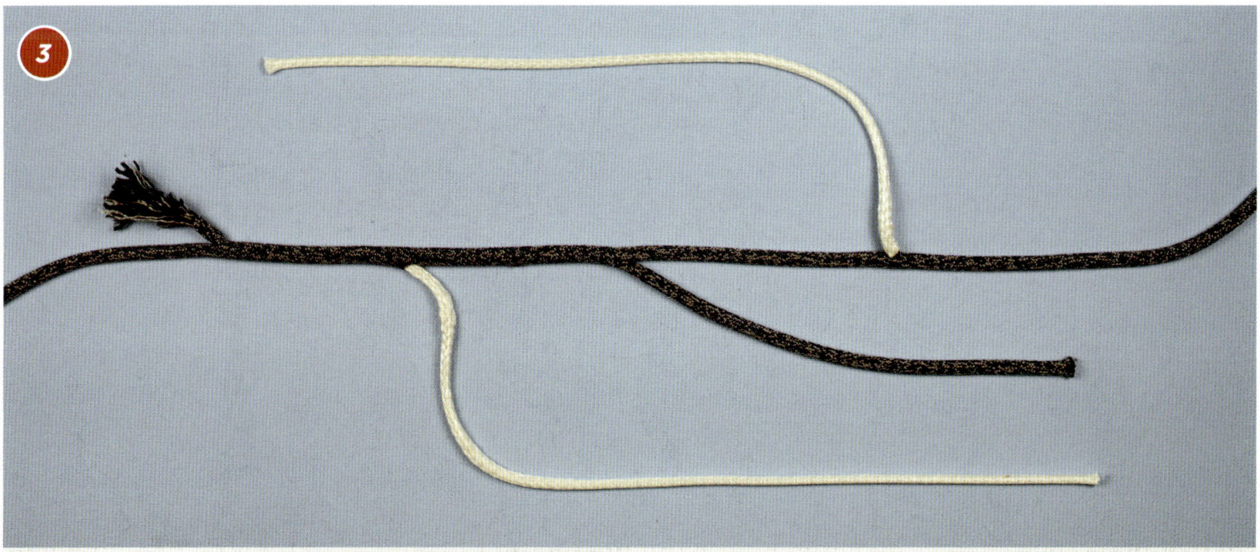

Pull the cover end through until it protrudes but a few centimeters.

Now the other cover end is pulled inside. Measure at the other side 20 times the rope's diameter from the spot where the core emerges back toward the standing end of the rope, and insert the splicing needle at this spot. Push the splicing neeedle along between core and cover and bring it out again at a distance of 20 times the rope's diameter to the other side of the spot where the core emerges. In total, the splicing needle is pushed through over a distance of 40 times the rope's diameter. Thread the other core end onto the splicing needle the same way as before and pull it through.

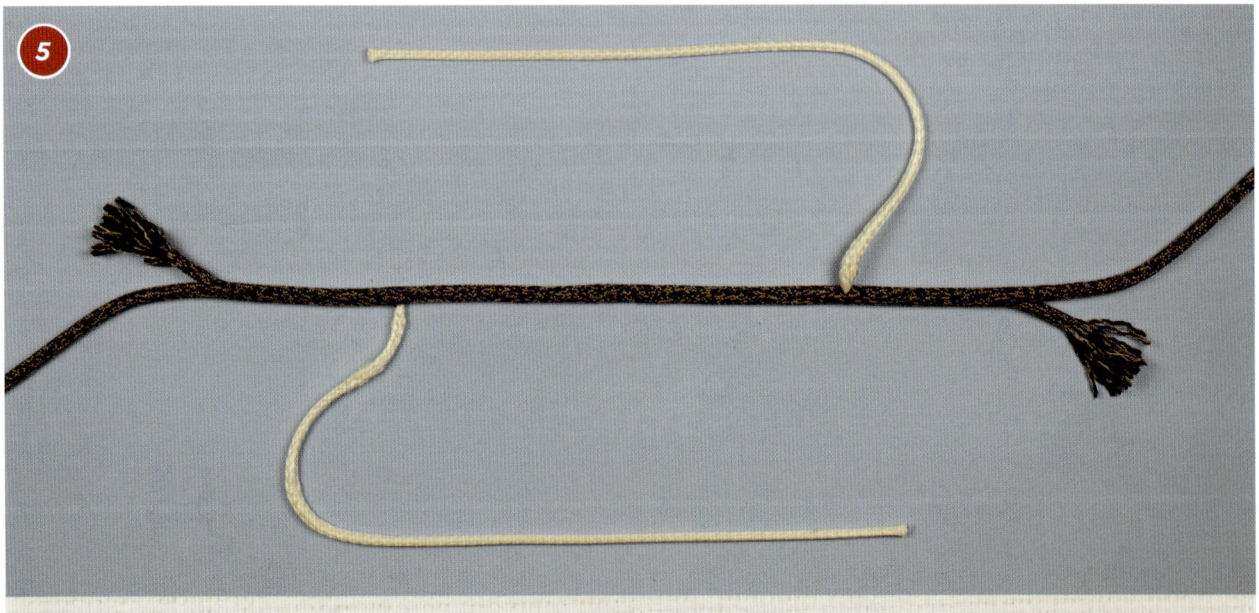

Both cover ends have been pulled through. Strongly pull at the protruding ends to tightly push the crossover spot together. Then firmly smooth back from the crossover spot toward both sides.

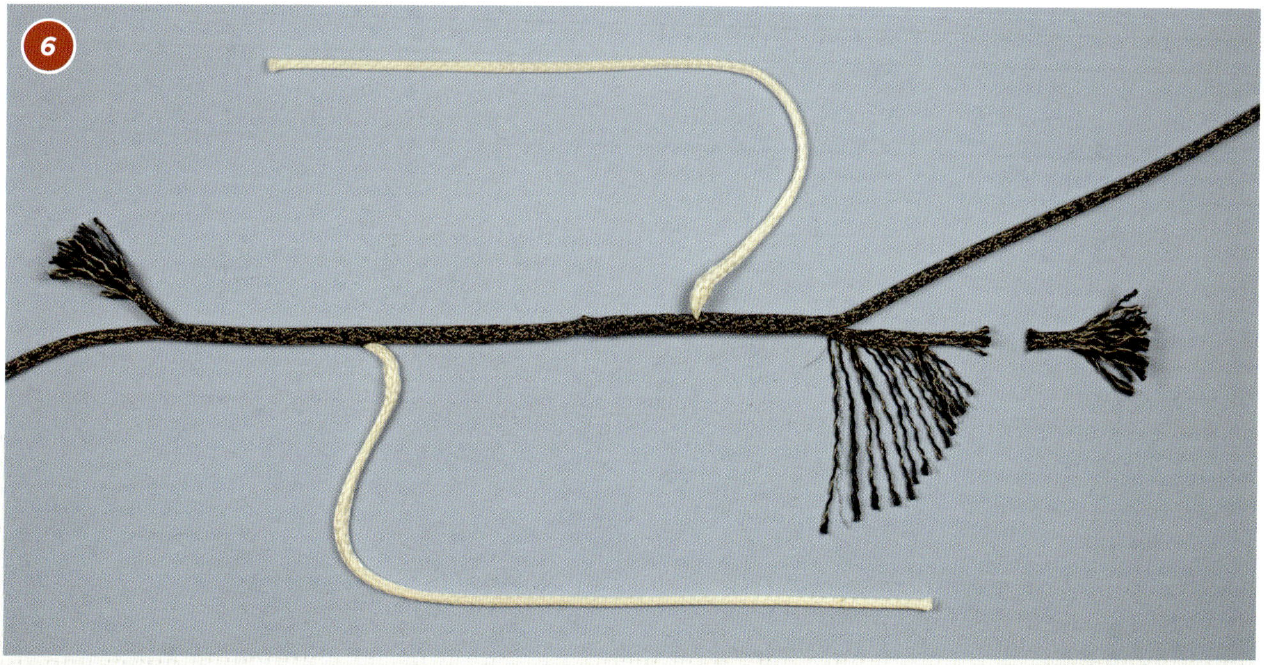

Now the cover ends are tapered. Hold one of the cover ends at exactly the spot where it emerges, and pull it out as far as possible. Cut it off at the spot where you are holding it. Taper the cover end by pulling out single yarns one behind the other and cut them off flush. More details about tapering can be found in the chapter "Basic Splicing Skills" on page 20.

Milk back firmly from the crossover spot, and the cover end diappears inside. Taper the other cover end in exactly the same way.

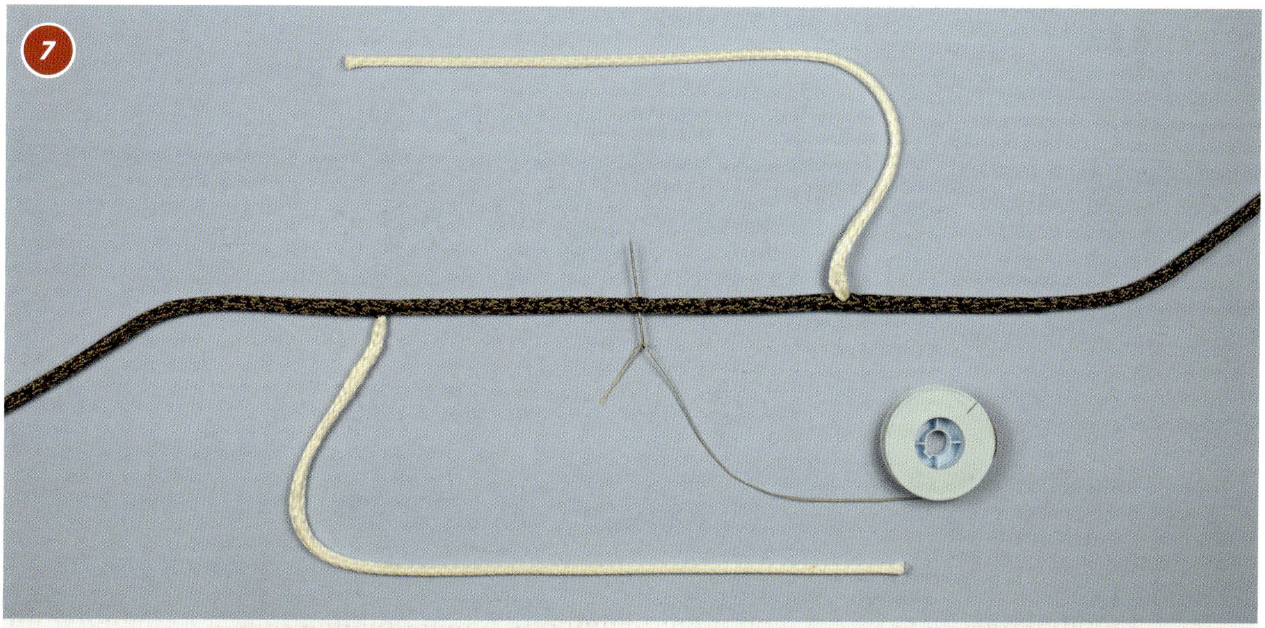

Both cover ends have disappeared inside the rope. Before taking care of the protruding core ends, the crossover spot is secured with a short length of whipping twine and several closely spaced stitches from one side of the crossover to the other.

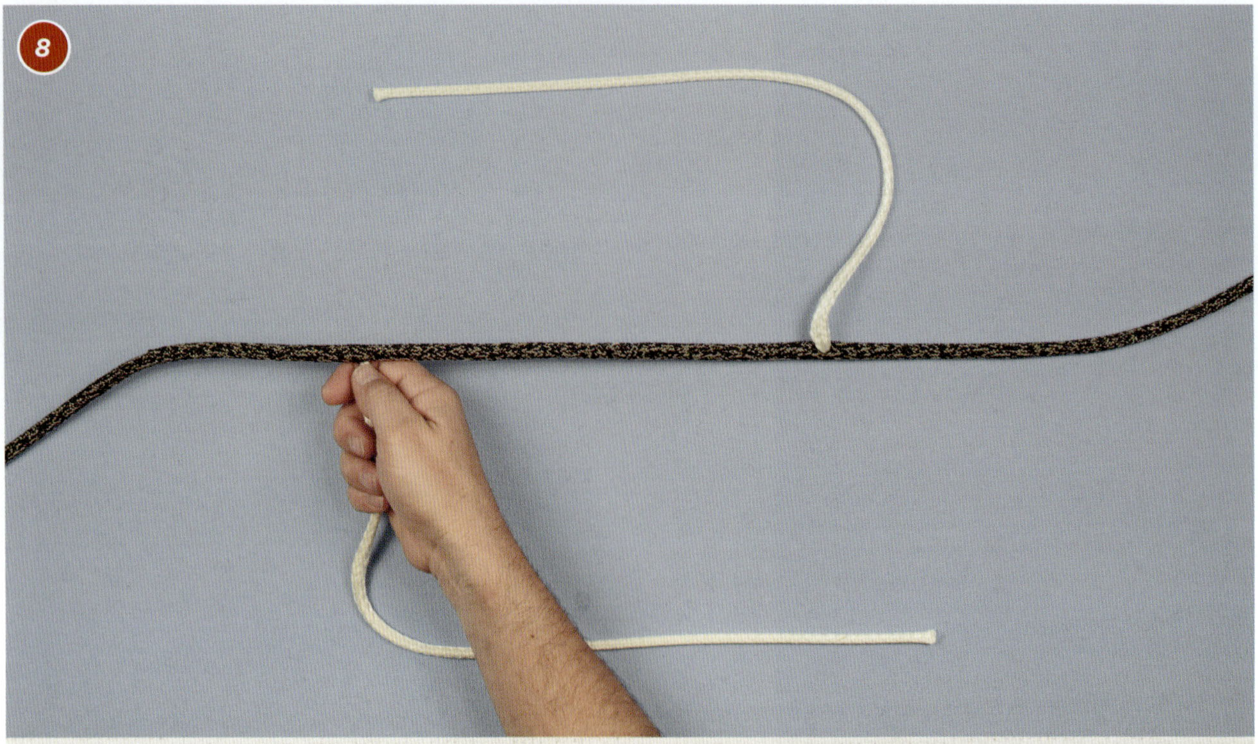

Now the core ends are cut and tapered. Grab one of the core ends at exactly the spot where it emerges.

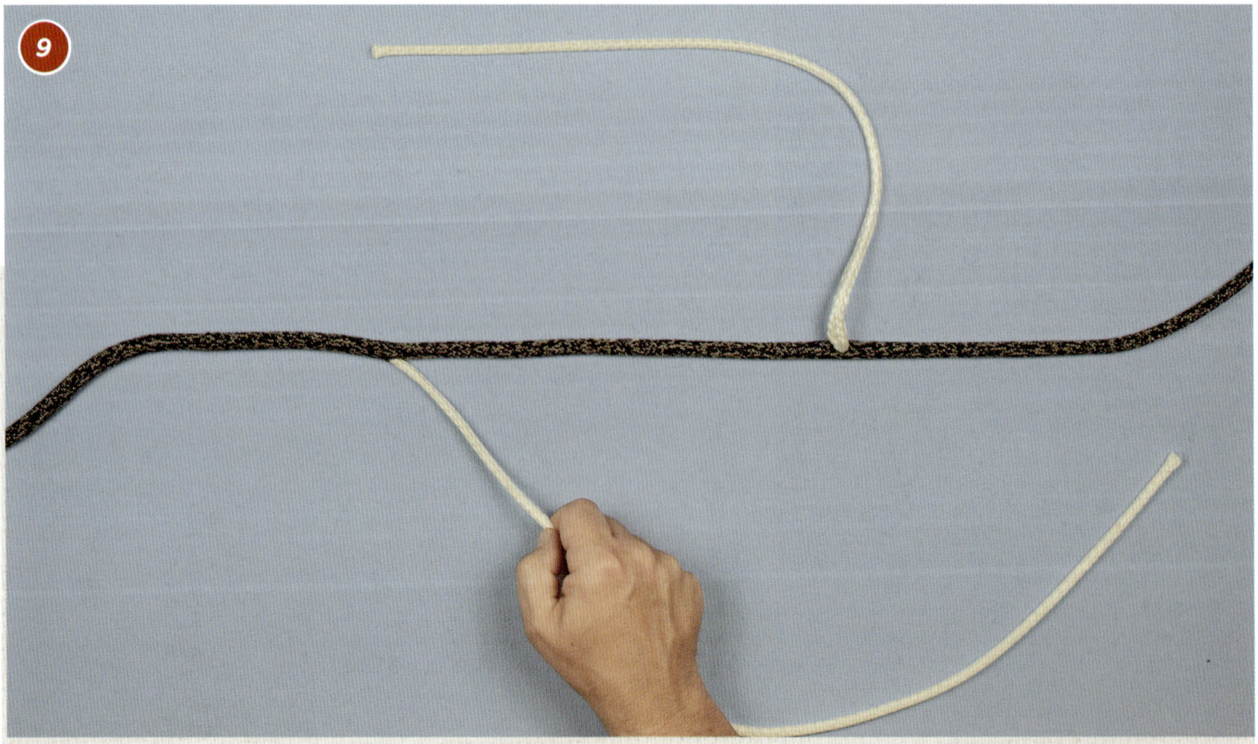

Pull the core end out over a distance of about 20 times the rope's diameter.

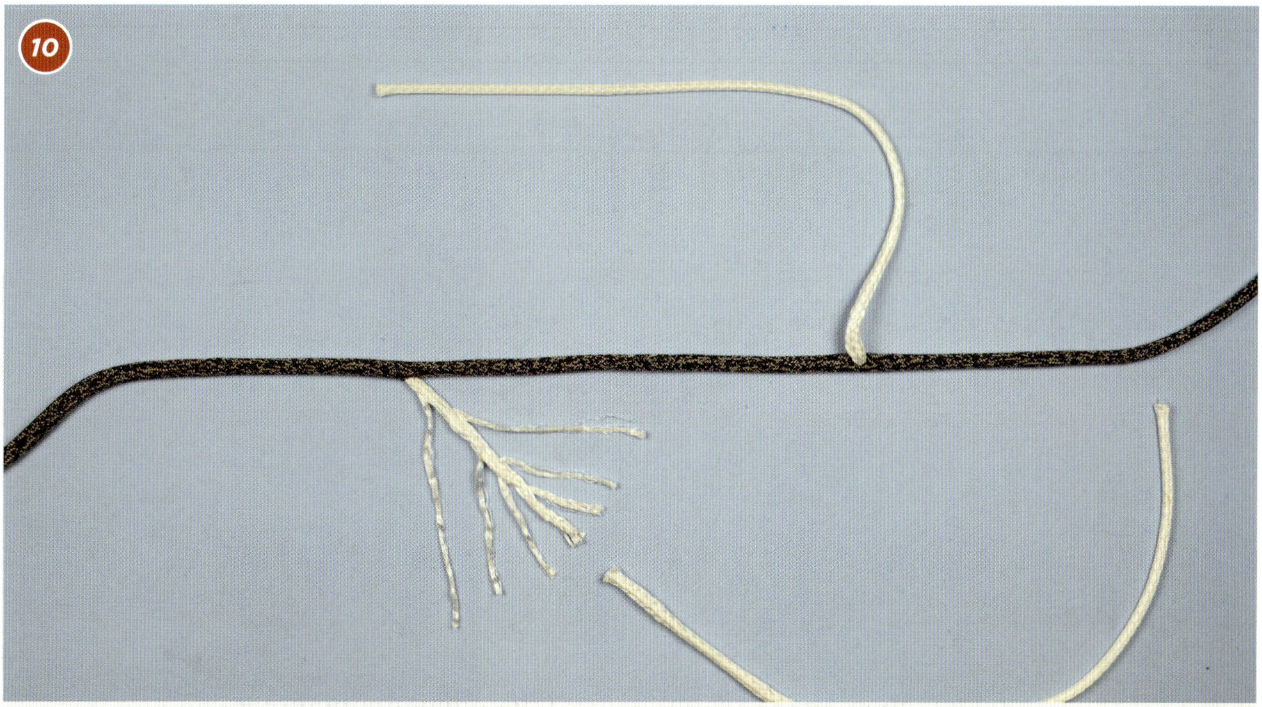

Cut the core end off at exactly the spot where you were holding it. Taper the end by pulling out strands at regular intervals, as shown. Cut those strands off flush.

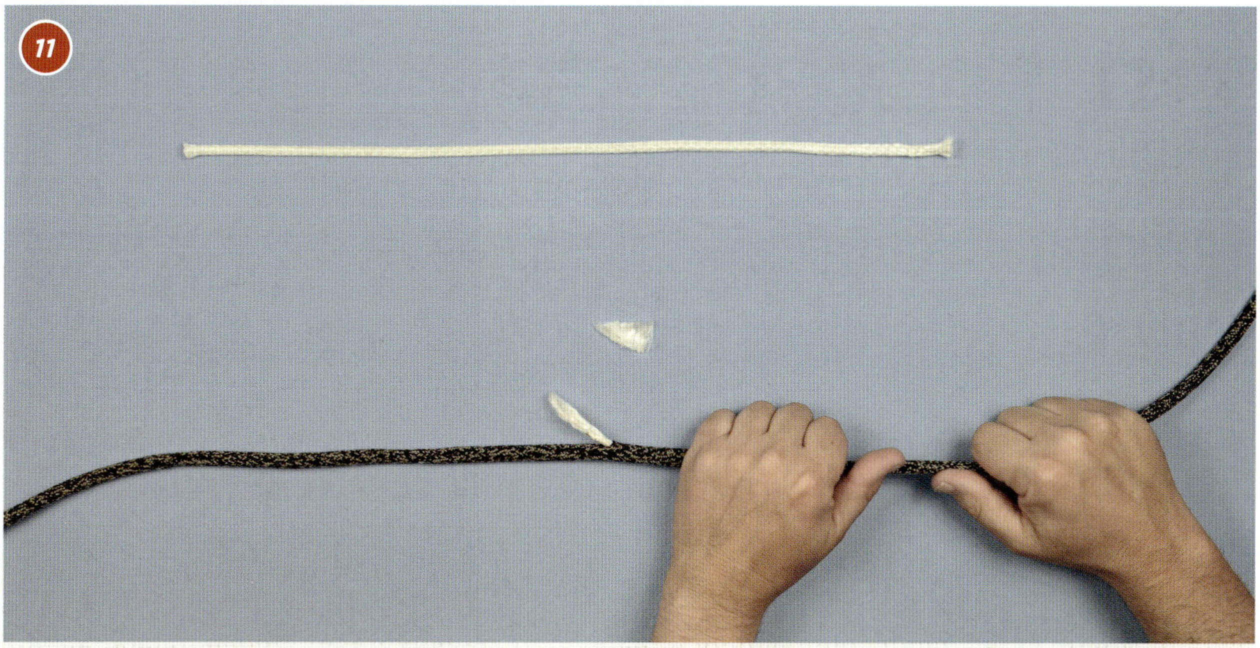

Cut the very end of the core at a sharp angle and milk back from the crossover spot. The core end disappears inside. Do exactly the same with the other core end. The tapers of the core and cover ends now overlap inside the rope and result in a uniform rope diameter along the whole end-to-end splice.

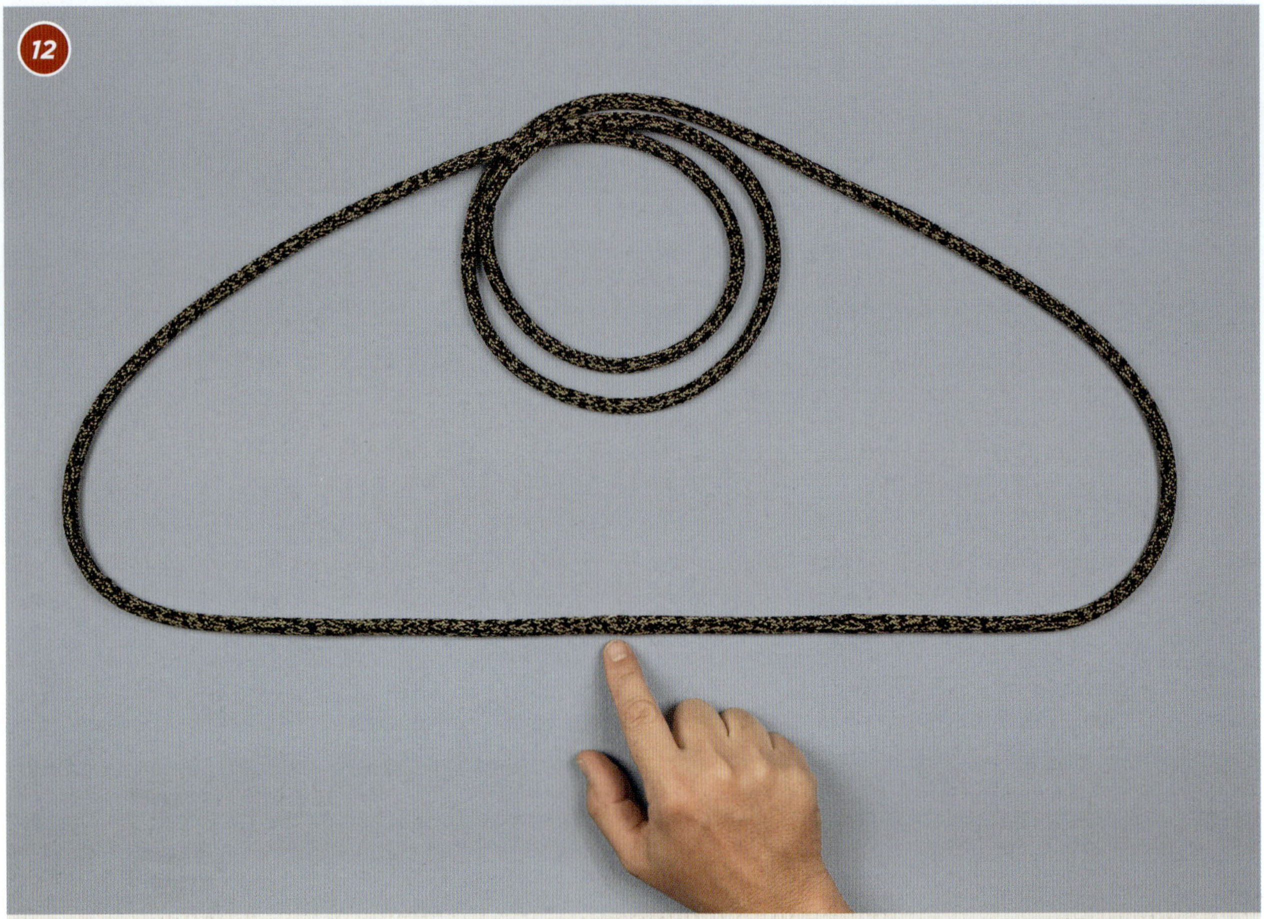

The finished end-to-end splice without any increase in diameter

2.6 TAPERING AND STRIPPING THE COVER OF ROPES WITH HIGH-MODULUS FIBER CORE

Tapering of sheets and halyards of double -braided construction (kernmantle ropes) with a coated core from high-modulus fiber such as Dyneema, Spectra, and other high-tech fibers by removing and stripping the cover at the tapered section. With the almost identical splicing technique, it is also possible to apply an additional cover to a sheet or halyard of single- or double -braided construction in order to improve holding power and protection in rope stoppers or on winches.

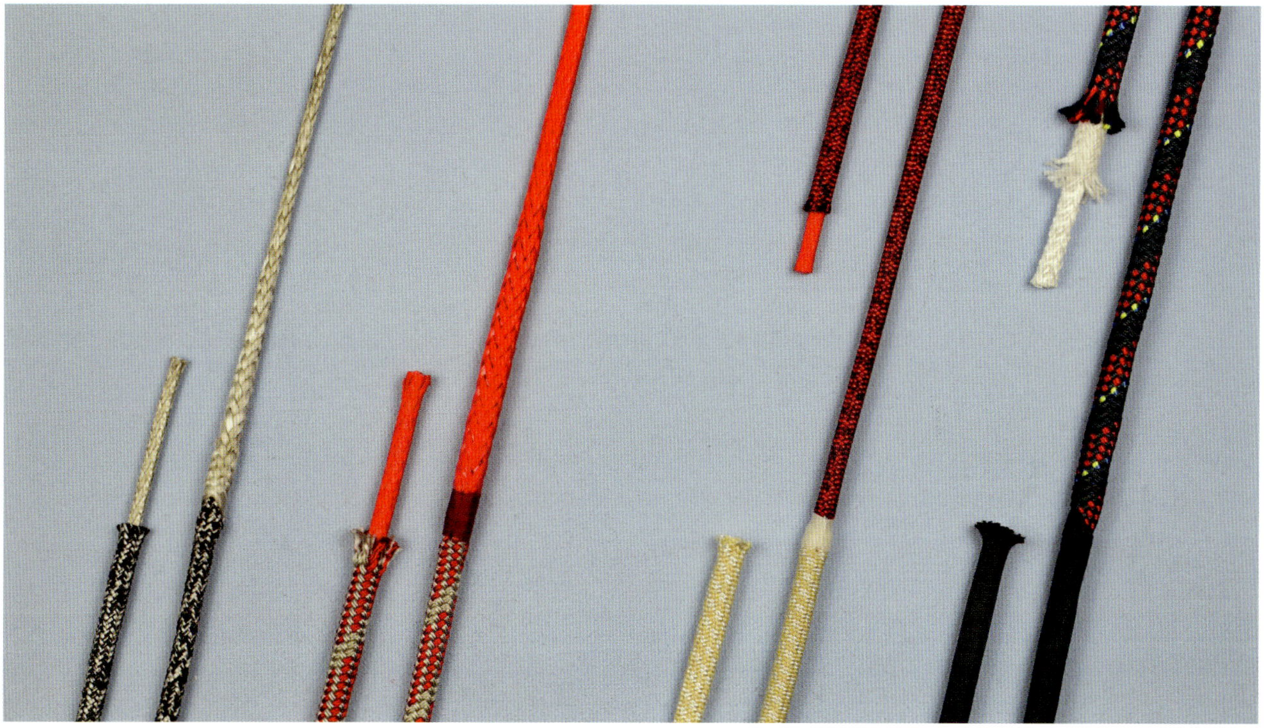

Tapering of spinnaker or gennaker sheets typically serves to reduce weight, but other lines such as halyards, topping lifts, downhauls, and other control lines can be equally tapered to save weight and reduce windage. If the tapered section of a line runs through purchase systems, the blocks can also be chosen in a smaller size, again saving weight, windage, and also cost. Ropes to be tapered must have a specially coated core from high-modulus fiber that is meant to be used without the cover.

Splicing an additional cover onto a section of a rope, on the other hand, is done to give better protection against wear, improve grip and heat resistance for use on winches, and generally prolong a rope's lifes pan. Quite often, halyards of conventional fiber material are being replaced with high-modulus fiber ropes. To maintain the same breaking strength, the new high-modulus fiber halyard can be chosen in a thinner diameter, but the extant halyard stoppers may have problems gripping the thinner line sufficiently. This is a typical case where splicing an additional cover to the section that is held in stoppers (all reefing points included) provides a sound solution. The spliced-on cover must stretch at least 1 meter / 3 ft. from the stopper toward the load-bearing side.

TAPERING

START

VIDEO 2.6

Construction: double braided (kernmantle rope)
Material: core: high-modulus fiber (e.g., coated Dyneema, Spectra), cover: polyester or mixed fiber
Difficulty: easy
Splicing tools: splicing needle or fid

1

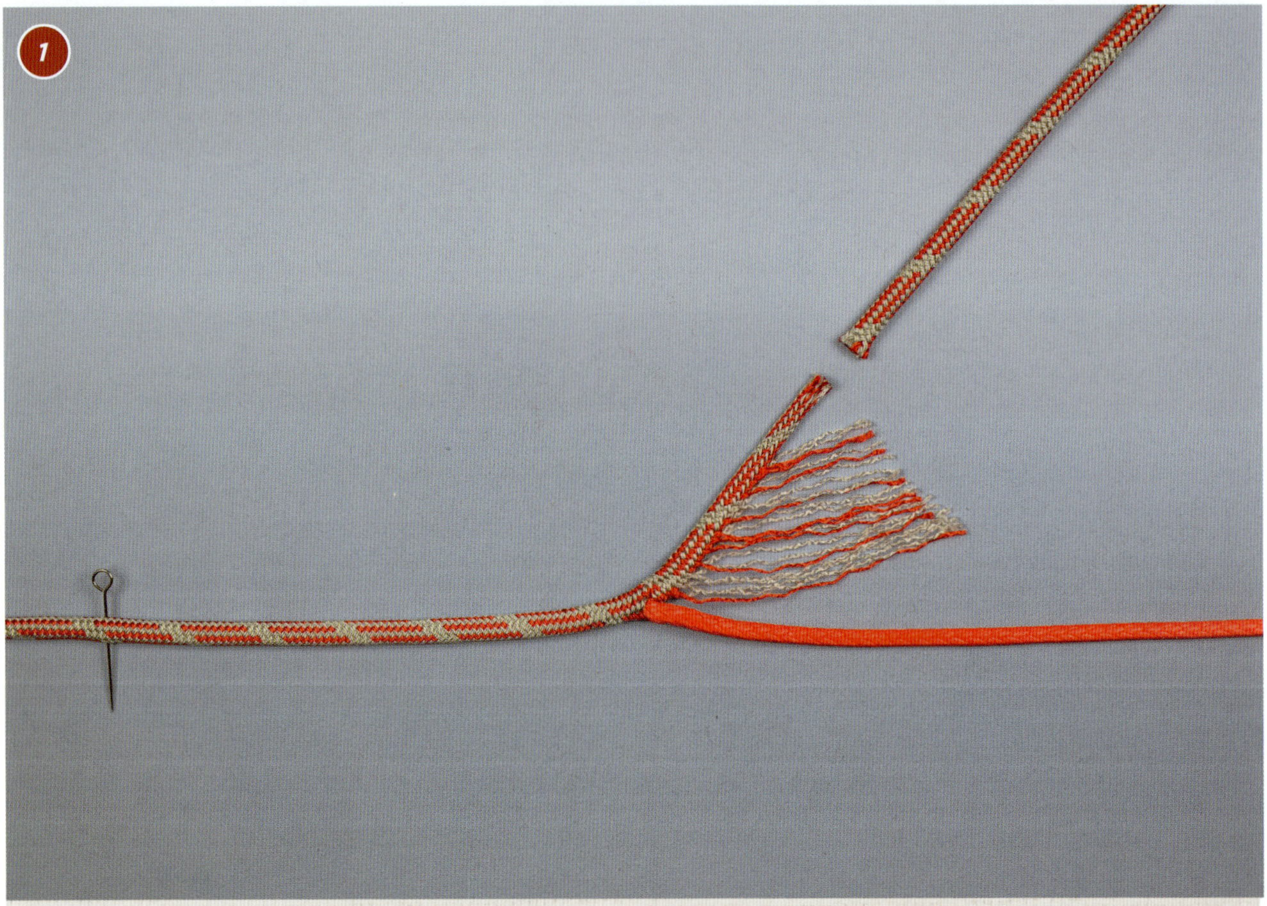

Extract the core at the spot where you want the taper to start. Secure the rope against any slippage of the core against the cover by piercing a small needle through the middle of the rope, a little distant from where the core was extracted. From the spot where the core was extracted, measure a length of 12 times the rope's diameter along the hollow cover end and cut the cover end off at this length. Taper the short cover end, as shown, by pulling out about half of its yarns one behind the other. All details about tapering can be found in the chapter "Basic Splicing Skills" on page 20.

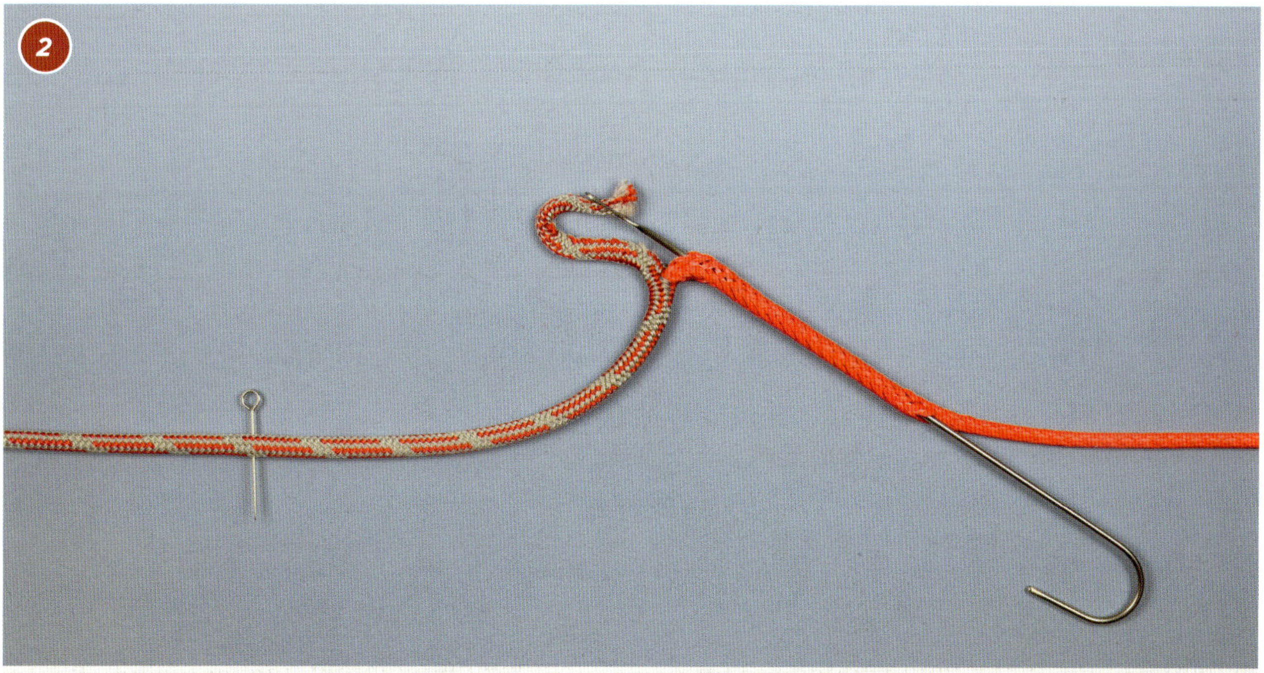

Insert the splicing needle at a sufficient distance into the core and push it along the inside of the core, toward the spot where the core emerges from the cover. Push the cover back a short distance of only about one rope diameter (here, 10 mm) and exit at exactly this spot with the tip of the splicing needle. Thread the tapered cover end onto the splicing needle and pull it through.

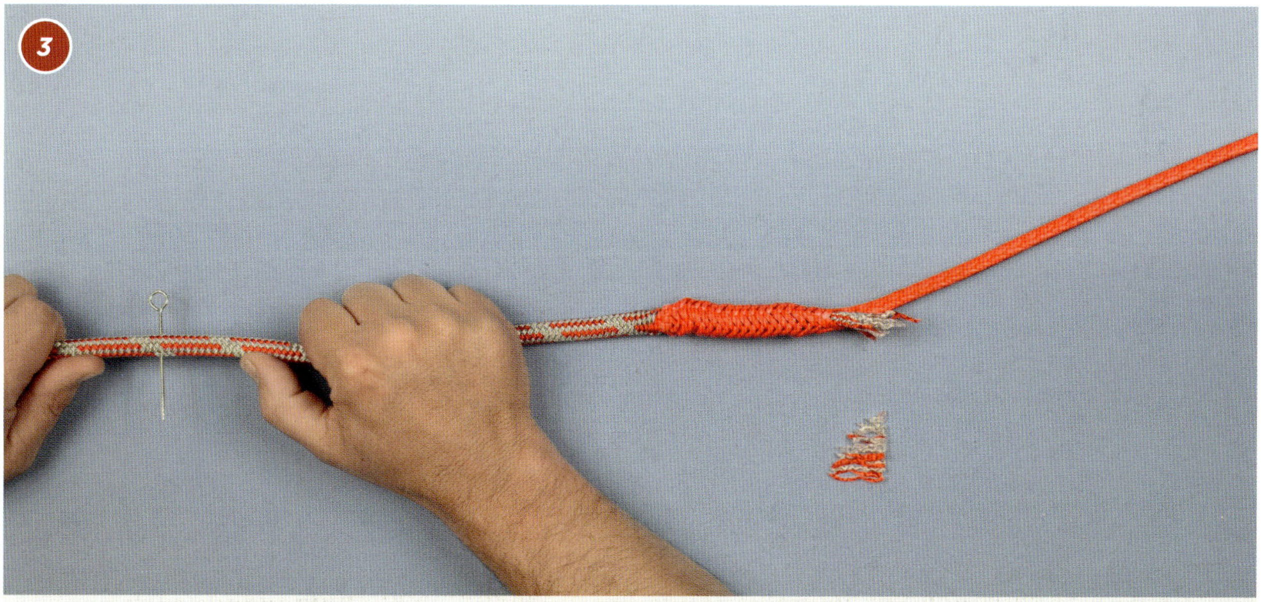

Give the protruding cover end a good pull, pushing back the core toward the crossover spot. Then cut the very end of the cover at a sharp angle to complete its taper. Firmly milk back the cover from the spot where the small needle has been pierced through the rope. This way, the crossover spot is very tightly pushed together. Milk back farther over the crossover and the exposed core, and the cover end disappears inside.

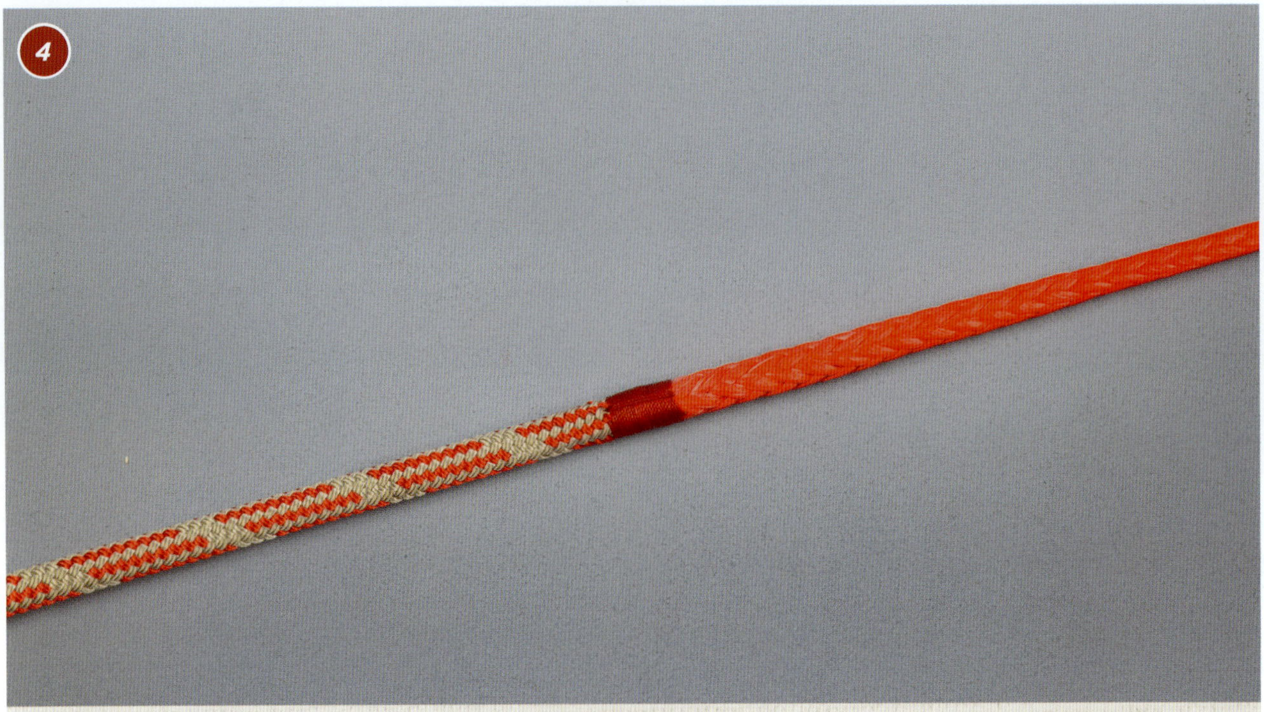

Secure the crossover spot with a palm-and-needle whipping (stitched whipping). Alternatively, the crossover can also be secured by just a few closely spaced stitches extending to both sides of it.

SPLICING AN ADDITIONAL PROTECTIVE COVER ONTO SHEETS AND HALYARDS

START VIDEO 2.6.1

Construction:	double braided (kernmantle rope)
Material:	high-modulus fiber (e.g., coated Dyneema, Spectra), cover: polyester or mixed fiber
Difficulty:	easy
Splicing tools:	splicing needle or fid

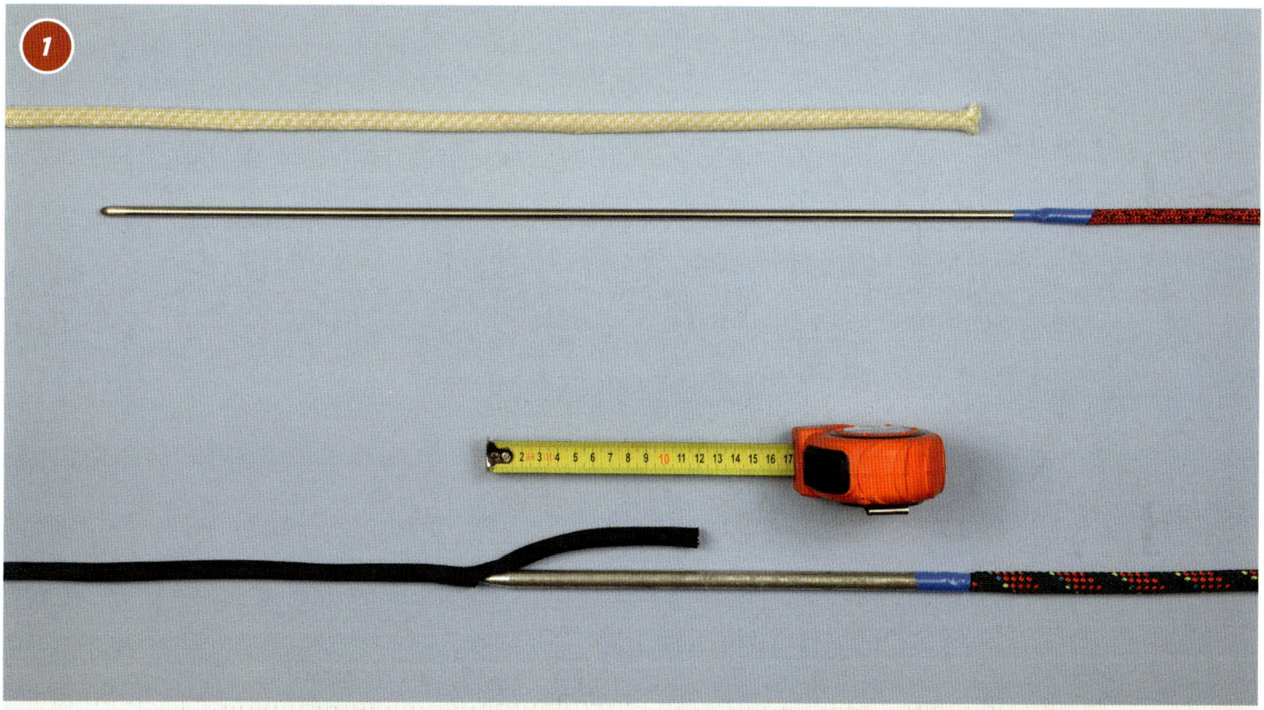

Attach the end of the rope to a fid or any kind of rod with a rounded tip. When splicing a hollow cover of any significant length onto a rope, a longer piece of rod of at least 60 cm / 2 ft. is advantageous in order to insert the rope more easily and quickly into the cover braid (*top*).

Measure a distance of 12 times the rope's diameter (here, 12 x 10 mm = 12 cm) from the end of the hollow cover braid. Widen an opening at this spot and insert the fid.

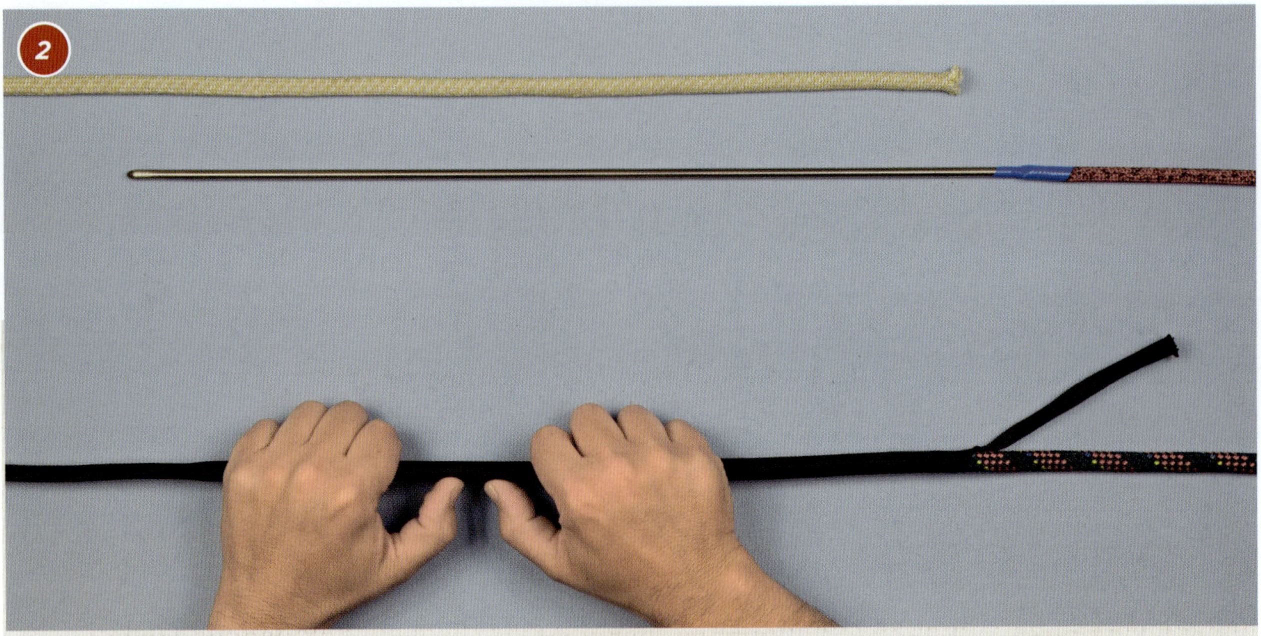

Push the fid with the line attached along the inside, by first holding the cover where the rear end of the fid is, and pushing the hollow cover over the tip of the fid as far as possible, then holding the cover where the tip of the fid is on the inside, and smoothing back the bunched cover over the length of the fid and farther down the rope. That way, the fid is pushed along on the inside step by step until the desired length of the additional cover is reached.

Taper the cover end by pulling out about half of its yarns one behind the other and insert the splicing needle at a sufficient distance to pull the cover end inside. Push the splicing needle along the inside between the core, including the intersheath if present, and the cover of the rope. Exit with the splicing needle at exactly the entry point of the rope into the additional cover.

Alternatively, the rope can also be inserted directly from the end of the additional cover braid (top).

4

Thread the end of the tapered cover onto the splicing needle and pull it through. Cut the very end at a sharp angle and milk back from the crossover point. The cover end disappears inside the rope.

5

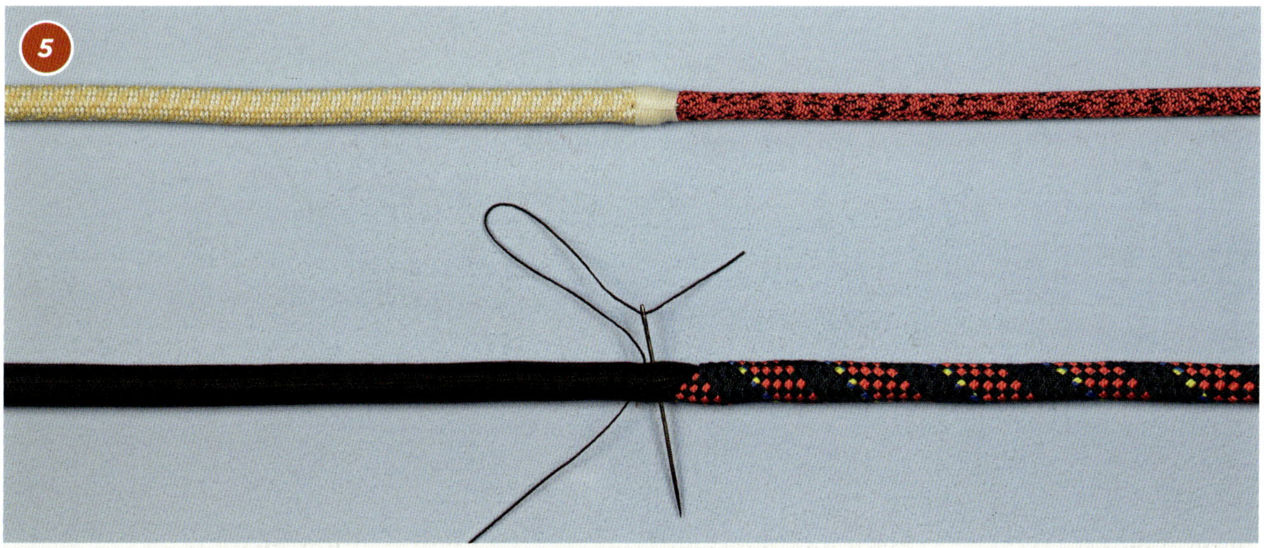

If the cover end is spliced into the rope (*bottom*), a few closely spaced stitches along the crossover point are enough to secure it. If the rope has been inserted directly from the end of the additional cover braid, however (*top*), it is better to secure the end with a palm-and-needle whipping (stitched whipping).

Firmly and repeatedly milk over the entire length of the additional cover braid, starting at the crossover point, until there is absolutely no slack left in the outer cover. Usually the additional cover will reach from one end of the rope to a desired length where it terminates with the spliced crossover or a whipping. In case the additional cover is to be applied only over a section somewhere in the middle of the rope, the other end can be equally spliced with a crossover or secured by a whipping.

TECHNICAL CHARACTERISTICS AND COMPARISON DATA ON FIBERS USED IN MODERN ROPES

Overview on the most-important data and attributes of fibers commonly used in modern cordage. Note the fundamental difference between conventinal and high-modulus fibers regarding both strength and working stretch but also knot tenacity.

	CONVENTIONAL FIBERS			HIGH-MODULUS FIBERS			
Material	Polyester	Polyamide	Polypropylene	HMPE		LCP	PBO
Trade name	Trevira, Dacron	Nylon, Cordura	Hempex, Nelson, etc.	Dyneema SK 78	Dyneema SK 99	Vectran	Zylon
Tensile strength —8 mm rope in daN (approx. kg)	1,650	1,600	900	5,300	7,000	5,000	8,400
Working stretch	5%	12%	9%	0.9%	0.9%	0.8%	0.5%
Breaking stretch, typical comparison	13%	22%	16%	3.8%	3.8%	3.3%	2.5%
Specific weight (g/cm^3)	1.38	1.14	0.91	0.96	0.96	1.41	1.52
Melting point	260°C	215°C	170°C	140°C	140°C	330°C	650°C
	500°F	419°F	338°F	284°F	284°F	626°F	1,202°F (charred)
Knot tenacity Percentage of BRL typical comparison data	60%	60%	60%	35%	35%	33%	40%
UV resistance	very high	high	medium	high	high	very low	very low
Water absorption	low	high	none	none	none	very low	very low
Abrasion resistance	very high	very high	medium	high	high	high	low

RECOMMENDED ROPE DIAMETER FOR SHEETS AND HALYARDS

This table may serve as a point of reference for dimensioning the running rigging aboard a cruising or performance yacht. The values of boat length, displacement, and sail area may be interpolated to the best approximation.

Boat length in feet	20-26	30	33	36	40	45	52	60
Displacement in tons	1.5-2.5	4	5-6	6.5-7.5	8-10	11.5-13	15-17	19-24
Sail area in square feet								
Main sail	100–215	215–270	270–300	320–390	400–450	480–620	590–650	750–970
Foresail, genoa	100–160	270	270–320	320–380	430–460	510–620	640–750	860–1070
Spinnaker, gennaker	270-540	540-640	590-800	860-1290	1290-1500	1600-1800	1800-2150	2500-3000

RECOMMENDED DIAMETER IN MM

Mainsail halyard, conventional	8–10	10–12	12	12–14	14	16	18	20
Mainsail halyard, high modulus	6–8	10	10	10–12	12	14	16	16–18
Mainsail sheet, conventional	8–10	10 -12	12	12–14	12–14	14	16	18
Mainsail sheet, high-modulus fiber	8	10	10	10–12	12	12	14	16
Genoa halyard, conventional	8–10	10–12	12	12–14	14	14–16	18	20
Genoa halyard, high-modulus fiber	6–8	10	10	10–12	12	12–14	16	18
Genoa sheet, conventional	8–10	12	12–14	14	14–16	16	18–20	20
Genoa sheet, high-modulus fiber	8	10	10-12	12	12-14	14	16-18	16-18
Spinnaker/ gennaker halyard, conventional	8–10	10	12	12–14	12–14	14	14–16	18
Spinnaker/ gennaker halyard, high-modulus fiber	6–8	8–10	10	10–12	12	12–14	14	16
Spinnaker/ gennaker sheet, conventional	8	10	10–12	12	12–14	14	14–16	18
Spinnaker/gennaker sheet, high-modulus fiber	6–8	8–10	10	10–12	12	12–14	14	16
Mooring line	10–12	12–14	14–16	16	16–18	16–18	20	24
Anchor rode	12	12–14	14–16	16	16–18	18	20	24

Conventional cordage (blue) — High-modulus cordage (yellow)

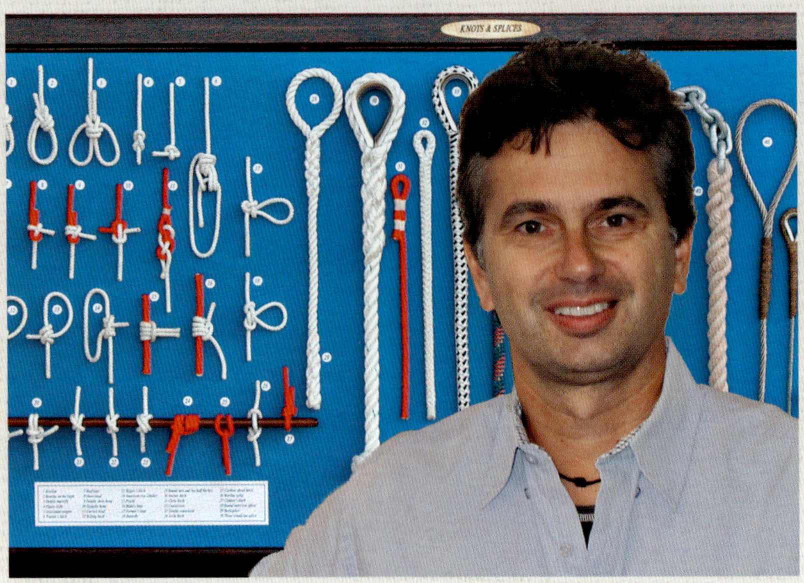

EGMONT M. FRIEDL is an expert on splicing, ropes, and rigging and holds a professional license for the 200-ton Yachtmaster. He teaches seminars on splicing worldwide and has worked in boatbuilding in California, Maryland, England, Italy, and Germany. www.emf-marine.com

TABLE OF SPLICING LENGTHS

Splicing lengths are given as multiples of rope diameter. This table allows for quick reference. Lengths are rounded off to 0.5 cm.

Rope Diameter	6-times	8-times	12-times	14-times	20-times	24-times	48-times
3 mm	2.0 cm	2.5 cm	3.5 cm	4.0 cm	6.0 cm	7.0 cm	14.5 cm
4 mm	2.5 cm	3.0 cm	5.0 cm	5.5 cm	8.0 cm	9.5 cm	19.0 cm
5 mm	3.0 cm	4.0 cm	6.0 cm	7.0 cm	10.0 cm	12.0 cm	24.0 cm
6 mm	3.5 cm	5.0 cm	7.0 cm	8.5 cm	12.0 cm	14.5 cm	29.0 cm
8 mm	5.0 cm	6.5 cm	9.5 cm	11.0 cm	16.0 cm	19.0 cm	38.5 cm
10 mm	6.0 cm	8.0 cm	12.0 cm	14.0 cm	20.0 cm	24.0 cm	48.0 cm
12 mm	7.0 cm	9.5 cm	14.5 cm	17.0 cm	24.0 cm	29.0 cm	57.5 cm
14 mm	8.5 cm	11.0 cm	17.0 cm	19.5 cm	28.0 cm	33.5 cm	67.0 cm
16 mm	9.5 cm	13.0 cm	19.0 cm	22.5 cm	32.0 cm	38.5 cm	77.0 cm
18 mm	11.0 cm	14.5 cm	21.5 cm	25.0 cm	36.0 cm	43.0 cm	86.5 cm
20 mm	12.0 cm	16.0 cm	24.0 cm	28.0 cm	40.0 cm	48.0 cm	96.0 cm
24 mm	14.5 cm	19.0 cm	29.0 cm	33.5 cm	48.0 cm	57.5 cm	115.0 cm